Enzyme Mathematics

STUDIES IN MATHEMATICS AND ITS APPLICATIONS

VOLUME 10

NORTH-HOLLAND PUBLISHING COMPANY – AMSTERDAM • NEW YORK • OXFORD

ENZYME MATHEMATICS

JEAN-PIERRE KERNEVEZ
Université de Technologie de Compiègne

1980

NORTH-HOLLAND PUBLISHING COMPANY – AMSTERDAM • NEW YORK • OXFORD

ISBN 0 444 86122 x

Publishers:
NORTH-HOLLAND PUBLISHING COMPANY
AMSTERDAM • NEW YORK • OXFORD

Sole distributors for the U.S.A. and Canada:
ELSEVIER NORTH-HOLLAND, INC.
52 VANDERBILT AVENUE
NEW YORK, N.Y. 10017

Library of Congress Cataloging in Publication Data

Kernevez, Jean Pierre.
Enzyme mathematics.

(Studies in mathematics and its applications ; v. 10)
Includes bibliographies.
1. Immobilized enzymes--Mathematical models.
I. Title. II. Series.
QP601.K36 574.19'25'0724 80-22449
ISBN 0-444-86122-X

PRINTED IN THE NETHERLANDS

PREFACE

In this book, the main goal is to present mathematical models for enzymes immobilized within membranes, to show how these models represent the biochemical phenomena, and to derive some interesting conclusions from these studies.

In this preface we first want to give very general indications on what enzymes are and how they behave. Then we shall give some details on the contents of the book.

Enzymes are molecules which catalyze the biochemical reactions in the metabolic pathways of living organisms, whence their importance. Most of the biochemical reactions are catalyzed by enzymes, each enzyme being highly specific for one type of reaction. But in living cells reaction interacts with diffusion because enzymes act within structured systems: either bound to cell organelles such as mitochondria, or linked to membrane structures where they catalyze metabolic reactions taking a part in the transfer of metabolites, or embedded within the cytoplasm, where the viscosity is high and the convection effects negligible. On the other hand, the complexity of living cells is such that it is difficult to model the whole system, where so many phenomena play a role. There are many enzyme reactions, many kinds of transport, by diffusion, electrical migration, convection or other active processes. Whence the idea of embedding enzyme molecules within artificial membranes in order to study the interaction of enzyme reaction and diffusion in a well defined context. Such membranes, where enzymes are physically confined and localized in a certain region of space with retention of their catalytic activities, and where enzymes can be used repeatedly and continuously, are called immobilized enzyme systems.

In these enzymatically active artificial membranes the interaction of diffusion and reaction produces macroscopic spatial, temporal, or functional structures in a self-organized way. The subject of this book is to study these phenomena. More precisely this interaction can give rise to concentration profiles (Chapter 1), active transport (Chapter 2), multiple steady states and hysteresis (Chapters 3 and 4), pattern formation (Chapters 5 and 6), oscillations and wave front propagation (Chapter 7). Biological phenomena such as active transport across membranes, short term memory, pattern formation in embryos, biological clocks, and transmission of information in living systems may have a similar origin.

Each chapter is devoted to the description of a precise system, together with the numerical and mathematical methods employed for its analysis. The mathematical technicalities are postponed until the last sections of each chapter, so that non-mathematicians may read the first sections, dealing with the description of the phenomena, the numerical methods employed for their simulation, and the results of these numerical simulations.

The first chapter is an introduction to enzyme kinetics and enzyme membranes. As a first example we present a simple case, represented by the equation:

$$(0.1)\quad \begin{cases} s_t - s_{xx} + \sigma s/(1 + |s|) = 0, \quad t > 0, \quad 0 < x < 1, \\ \text{with boundary conditions} \\ s(0,t) = \alpha \text{ and } s(1,t) = \beta \\ \text{and initial condition} \\ s(x,0) = 0. \end{cases}$$

Here σ, α, and β are positive parameters and $s = s(x,t)$ denotes the concentration of substrate at (x,t) in a one-dimensional membrane.

It is the occasion to point out the important notion of concentration profile, and to show how simple the numerical algorithms may be, whose trivial generalizations enable one to solve more complex systems numerically. We also give a method for proving existence and uniqueness of a positive solution for the evolution equations which describe such diffusion-reaction systems.

The second chapter describes a "glucose pump" which has the property of transporting glucose from one side of a membrane to the other against the concentration gradient, i.e. from a region of low concentration to a region of high concentration.

Chapter 3 deals with the possibility of multiple steady states in a system with hysteresis, governed by the equations:

$$(0.2)\quad \begin{cases} -\dfrac{d^2s}{dx^2} + \sigma \dfrac{s}{1 + s + ks^2} = 0, & 0 < x < 1, \\ s(0) = s(1) = \lambda. \end{cases}$$

We study the S-shaped curve of solutions as the parameter λ varies, both numerically (by the Kubicek method of continuation) and mathematically (by applying the implicit function theorem and results of Crandall and Rabinowitz). Lastly, we justify the consideration of sequential steady states as the parameter varies by proving the so-called quasi-steady state hypothesis.

Assemblages of cells are presented in Chapter 4 and are the occasion to discuss bifurcation phenomena and numerical methods for exploring a whole connected set of solutions including bifurcating branches. Here we mainly employ H.B. Keller's methods. Finally we analyse the effect of imperfections.

Chapters 5 and 6 are devoted to pattern formation in enzyme systems. More precisely, chapter 5 deals with an assemblage of N cells, each cell containing an enzyme which catalyzes a reaction involving two reactants, S and A. These reactants are supplied by an external reservoir to each cell by diffusion across a membrane. In the absence of diffusion coupling between cells, there exists a stable steady state, with uniform concentrations of S and A in all cells. But if we allow two neighboring cells to communicate by diffusion, this steady state becomes unstable and new stable steady states arise, which are no longer uniform, but with different concentrations of S and A in the cells. Chapter 5 thoroughly studies this bifurcation of structured steady states from the branch of uniform steady states.

Chapter 6 studies the same kind of behavior, that is diffusion-driven instability, in one or two dimensional spatial domains. Here the governing equations are:

$$(0.3)\quad \begin{cases} s_t - \Delta s + \lambda[R(s,a) - (s_o - s)] = 0 \quad \text{in } \Omega \\ a_t - \beta\Delta a + \lambda[R(s,a) - \alpha(a_o - a)] = 0 \\ R(s,a) = \rho as/(1 + |s| + ks^2) \\ \text{with zero-flux boundary conditions,} \end{cases}$$

where Ω is a bounded domain in R^n ($n = 1,2$, or 3), more precisely an enzyme membrane, and the conditions on the parameters s_o, a_o, α, ρ, and β are such that, in particular, the following holds:

(0.4) there is exactly one solution, $(\tilde{s},\tilde{a})$, to the algebraic system,

$$(0.5)\quad \begin{cases} R(\tilde{s},\tilde{a}) - (s_o - \tilde{s}) = 0 \\ R(\tilde{s},\tilde{a}) - \alpha(a_o - \tilde{a}) = 0 \end{cases}$$

and

(0.6) $(\tilde{s},\tilde{a})$ is a stable node for the dynamical system:

$$(0.7)\quad \begin{cases} \dfrac{ds}{dt} + R(s,a) - (s_o - s) = 0 \\ \dfrac{da}{dt} + R(s,a) - \alpha(a_o - a) = 0 \end{cases}$$

It is plain that $(\tilde{s},\tilde{a})$ is a trivial steady-state solution of (0.3). By steady state solution of (0.3) we mean a solution to the stationary problem:

$$(0.8)\quad \begin{cases} -\Delta s + \lambda[R(s,a) - (s_o - s)] = 0 \quad \text{in } \Omega \\ -\beta\Delta a + \lambda[R(s,a) - \alpha(a_o - a)] = 0 \\ \text{with zero-flux boundary conditions.} \end{cases}$$

The important fact is that, as in the preceding chapter, this equilibrium solution $(\tilde{s},\tilde{a})$ may become unstable as λ crosses a critical value, and new spatially non-uniform stable steady state solutions arise.

We point out the similarity of the observed phenomena with morphogenesis in embryos. However by no means do we claim that our S-A system is responsible for pattern formation in embryos. However, the analogy is so striking and the possible consequences so important that this single phenomenon should justify our study of immobilized enzyme systems.

In chapter 7 we present experimental and numerical observations of oscillating enzyme systems, and numerical results about wave front propagation in the so-called S-A system.

At the end of each chapter we give references, admittedly incomplete, to literature on the topics discussed, our aim being to give an idea of the biochemical and biological background without being exhaustive, and to direct the reader to additional information on numerical and mathematical methods.

The intention of the book is two-fold: first to show some striking phenomena such as pattern formation, and secondly to present some numerical and mathematical methods in a simple form. The numerical methods described include finite difference and finite element analysis of reaction-diffusion problems, and procedures for solving bifurcation and nonlinear eigenvalue problems.

These numerical continuation procedures are two-step methods, as described by Kubicek and H.B. Keller: a predictor step employing Euler or Adams-Bashforth method, and a corrector step using Newton-Raphson method. The numerical methods for bifurcation of steady-states are those of H.B. Keller.

The mathematical methods deal with existence and uniqueness of a positive solution for the evolution equations governing our systems, and existence, multiplicity and stability of their steady state solutions.

We show how it is practical to use upper and lower solutions, as employed by Amann or Sattinger.

Special emphasis is put on analysis of a continuum of steady state solutions as some parameter varies, and bifurcation of steady state solutions, for which we mainly refer to Crandall and Rabinowitz, and H.B. Keller.

It is a pleasure to acknowledge the contribution of Daniel Thomas. I am much indebted to him for the model systems hereafter described. He provided not only their description, but also the equations, the exact form of the rate expressions, and, which is very important, the numerical values of the parameters. Moreover his enthousiasm was the motivation of this work.

I would like to thank Jacques-Louis Lions whose teaching and encouragements have had such a major influence on my work.

Thanks go to all those who contributed or are contributing to the numerical and mathematical work on these systems: H.T. Banks, B. Bunow, C.M. Brauner, M.C. Duban, D. Dubus, J. Henry, G. Joly, C. Milan, P. Penel, J.P. Puel, J.P. Quadrat, L. Tartar, B. Viot, and J.P. Yvon.

Finally, a special thanks goes to Cindy Moreau for her excellent typing of the manuscript and her fine draftsmanship on the figures.

Compiègne 1980 Jean-Pierre Kernevez

TABLE OF CONTENTS

CHAPTER 1

DIFFUSION AND REACTION IN AN ENZYME MEMBRANE

In this chapter we present our basic biochemical model, governed by the equation:

$$
(1.1)\quad \begin{cases} s_t - s_{xx} + \sigma s/(1 + |s|) = 0, \quad t > 0, \quad 0 < x < 1, \\ \text{with boundary conditions} \\ s(0,t) = \alpha \quad \text{and } s(1,t) = \beta \\ \text{and initial condition} \\ s(x,0) = 0. \end{cases}
$$

Here σ, α, and β are positive parameters and $s = s(x,t)$ denotes the concentration of substrate at (x,t) in a one-dimensional membrane.

The motivation for studying such immobilized enzyme systems will be the subject of section 1.1. Section 1.2 presents a brief introduction to enzyme kinetics. Artificial enzymatically active enzyme membranes are described in Section 1.3. Their modeling by partial differential equations is given in Section 1.4. The numerical solution of these equations is indicated in Section 1.5, together with numerical results. Finally, Section 1.6 studies the existence and uniqueness of a positive solution for the evolution problem (1.1), and the behavior of the concentration profile $s(\cdot,t)$ as $t \to +\infty$.

1.1 Motivation for Studying Artificial Enzyme Membranes

Enzymes are molecules which catalyze the biochemical reactions in the metabolic pathways of living organisms, whence their importance. Most of the biochemical reactions are catalyzed by enzymes, each enzyme being highly specific for one type of reaction. But in living cells reaction interacts with diffusion because enzymes act within structured systems: either bound to cell organelles such as mitochondria or linked to membrane structures where they catalyze metabolic reactions taking a part in the transfer of metabolites, or embedded within the cytoplasm, where the viscosity is high and the convection effects negligible. On the other hand, the complexity of living cells is such that it is difficult to model the whole system, where so many phenomena play a role. There are many enzyme reactions, many kinds of transport, by diffusion, electrical migration, convection or other active processes. Whence the idea of embedding enzyme molecules within artificial membranes in order to study the interaction of enzyme reaction and diffusion in a well defined context. Such membranes, where enzymes are physically confined and localized in a certain region of space with retention of their catalytic activities, and where enzymes can be used repeatedly and continuously, are called immobilized enzyme systems.

1.2 Enzyme Kinetics

Though enzyme reactions have been used for thousands of years in fermentations, it was not before the nineteenth century that the idea became clear that the biochemical reactions are not under the influence of some mysterious principle of Life, but are subject to the usual Law of Mass Action like other chemical reactions, and are catalyzed by enzymes. In 1835, Berzelius suggested that most of the biochemical reactions could occur under the influence of a new force which he calls "catalytic". In 1878, Kühne coined the term "enzyme" for this catalytic force, from a greek word meaning "in yeast". In 1902, Brown suggested that the enzyme E combines with its substrate, S, to form an intermediate complex, ES, which then decomposes into the product, P, and E:

$$(1.2) \qquad E + S \underset{k_2}{\overset{k_1}{\rightleftharpoons}} ES \xrightarrow{k_2} E + P.$$

This can be described in the "lock and key" theory, (Figure 1.1), where the structures and shapes of enzyme and substrate molecules explain the high specificity of enzymes to substrate type observed in enzyme reactions. Other important facts can be explained by Brown's scheme and the "lock and key" theory: the possibility of inhibition of enzyme reactions by substances, called inhibitors, which can take the place of S on the enzyme molecule; the enzyme concentration may be small with respect to the substrate concentration, since enzyme molecules are released unchanged to be used again.

In 1903, Henri [1] , and, ten years later, Michaelis and Menten [2], translated Brown's chemical scheme into a mathematical model, by using the Law of Mass Action:

$$(1.3)\quad \begin{cases} \dfrac{d[S]}{dt} = -k_1[E][S] + k_{-1}[ES], \quad [S](0) = S_o, \\ \dfrac{d[ES]}{dt} = -(k_{-1} + k_2)[ES] + k_1[E][S], \quad [ES](0) = 0, \\ \dfrac{d[P]}{dt} = k_2[ES], \quad [P](0) = 0, \\ [E] + [ES] = E_o. \end{cases}$$

Here [E], [S], [ES] and [P] are the concentrations of free enzyme, free substrate, enzyme-substrate complex and product.

The Henri-Michaelis-Menten derivation uses the so-called <u>quasi - equilibrium assumption</u>, where the reaction $E + S \rightleftharpoons ES$ is supposed to be alone and at equilibrium:

$$0 = -k_1[E][S] + k_{-1}[ES] = -k_1(E_o - [ES])[S] + k_{-1}[ES],$$

so that one finds:

$$(1.4)\quad \begin{cases} [ES] = E_o[S]/(K_S + [S]) \\ \dfrac{d[P]}{dt} = V_M[S]/(K_S + [S]) \end{cases}$$

with $K_S = k_{-1}/k_1$ and $V_M = k_2E_o$.

In 1913, Bodenstein points out that a less restrictive assumption than the quasi-equilibrium assumption is to suppose that $d[ES]/dt = 0$ (the so-called <u>quasi-steady-state assumption</u>). In 1925, Briggs and Haldane [3] use this assumption to write:

$$0 = -(k_{-1}+k_2)[ES] + k_1[E][S] = -(k_{-1}+k_2)[ES] + k_1(E_o-[ES])[S]$$

which still yields (1.4), but now the "Michaelis constant" K_S is given by:

(1.5) $K_S = (k_{-1} + k_2)/k_1,$

and the rate expression for S is found to be (by using the first equation in (1.3) and the first equation in (1.4)):

(1.6) $$r = -\frac{d[S]}{dt} = \frac{d[P]}{dt} = V_M[S]/(K_S + [S]).$$

In 1967, Heineken, Tsuchiya, and Aris [4] justify this modeling by using singular perturbations. Indeed, the equations for S and ES alone may be written:

$$\frac{d[S]}{dt} = -k_1(E_o - [ES])[S] + k_{-1}[ES], \quad [S](0) = S_o$$

$$\frac{d[ES]}{dt} = -(k_{-1}+k_2)[ES] + k_1(E_o - [ES])[S], \quad [ES](0) = 0$$

We nondimensionalize by setting: $s = [S]/K_S$, $c = [ES]/E_o$, $\varepsilon = E_o/K_S$.

Here and hereafter, K_S is defined by (1.5).

Then the differential equations for s and c take the form:

(1.7) $$\begin{cases} \frac{ds}{dt} = \varepsilon(k_{-1}c - k_1K_S\, s(1-c)) & s(0) = s_o \\ \frac{dc}{dt} = k_1K_S(s - c(1+s)) & c(0) = 0 \end{cases}$$

If we take the stand that the large time behavior is the important one, we employ a larger time scale, ε^{-1}, and define $T = \varepsilon t$. Then rewriting (1.7) in terms of this new variable, we find:

(1.8) $$\begin{cases} \frac{ds}{dT} = k_{-1}c - k_1K_S\, s(1-c), & s(0) = s_o \\ \varepsilon\,\frac{dc}{dT} = k_1K_S(s - c(1+s)), & c(0) = 0. \end{cases}$$

A naive formal approximation is found by writing $\varepsilon = 0$, to obtain:

(1.9) $$c = \frac{s}{1+s}\ , \quad \frac{ds}{dT} = -k_2\,\frac{s}{1+s}\ .$$

This is tantamount to making the quasi-steady-state hypothesis, since it gives:

$$(1.10) \qquad [ES] = E_o[S]/(K_S+[S]) \quad \text{and} \quad \frac{d[S]}{dt} = -V_M[S]/(K_S+[S]) = -\frac{d[P]}{dt}.$$

Another naive formal approximation valid for moderate time is obtained by setting $\varepsilon = 0$ in (1.7):

$$\frac{ds}{dt} = 0, \qquad \text{(whence } s(t) = s_o\text{)}$$

$$\frac{dc}{dt} = k_1K_S(s_o - c(1+s_o)), \qquad c(0) = 0,$$

whence:

$$c(t) = \frac{s_o}{1 + \varepsilon_o}(1 - e^{-k_1K_St}).$$

Thus we see that during a short interval of time, of the order of $1/\varepsilon$, s remains approximately constant while c rapidly varies from 0 to $s_o/(1 + s_o)$. The larger time behavior, which is usually the important one in biological contexts is described by (1.9), or, with different units (1.10). Thus we have seen that the Briggs-Haldane quasi-steady-state assumption is justified in the usual case where E_o is small with respect to K_S. More details about the singular perturbation analysis of enzyme kinetics may be found in ([5],[6]).

The same kind of quasi-steady-state assumption may be applied to the kinetics of inhibited reactions. A <u>competitive inhibitor</u> I competes with S for the active sites on the enzyme molecule, and we have the following scheme:

$$E + S \underset{k_{-1}}{\overset{k_1}{\rightleftharpoons}} ES \xrightarrow{k_2} E + P$$

$$E + I \underset{k'_{-1}}{\overset{k'_1}{\rightleftharpoons}} EI$$

with the corresponding equations:

$$(1.11)\quad \begin{cases} \dfrac{d[S]}{dt} = -k_1[E][S] + k_{-1}[ES] & [S](0) = S_o \\ \dfrac{d[ES]}{dt} = k_1[E][S] - (k_1 + k_2)[ES] & [ES](0) = 0 \\ \dfrac{d[EI]}{dt} = k'_1[E][I] - k'_{-1}[EI] & [EI](0) = 0 \\ \dfrac{d[P]}{dt} = k_2[ES] & [P](0) = 0 \end{cases}$$

The quasi-steady-state assumption can be written, in this case, $d[ES]/dt = d[EI]/dt = 0$, whence $[E] = K_S[ES]/[S]$ and $[EI] = (K_S/K_I)$ $[ES][I]/[S]$, where $K_I = (k'_{-1}/k'_1)$ is the so-called inhibition constant of E for I. Putting these expressions of [E] and [EI] in the last equation of (1.11), we obtain:

$$(1.12)\quad [ES] = \frac{E_o}{\dfrac{K_S}{S} + 1 + \dfrac{K_S}{K_I}\dfrac{[I]}{[S]}} = E_o \frac{[S]}{K_S(1 + \dfrac{[I]}{K_I}) + [S]} .$$

Thus we find the following rate expressions for S and P:

$$(1.13)\quad \frac{d[P]}{dt} = -\frac{d[S]}{dt} = V_M \frac{[S]}{K_S(1 + \dfrac{[I]}{K_I}) + [S]} .$$

If we compare equations (1.10) and (1.12)-(1.13), we observe that a competitive inhibitor increases the apparent Michaelis constant of the enzyme for the substrate. It is worth noting the significance of these two constants V_M and K_S which appear in equations (1.9):

$$(1.14)\quad V_M = k_2E_o \quad \text{and} \quad K_S = (k_{-1} + k_2)/k_1 .$$

V_M is the maximal value of the reaction rate. K_S is the value of [S] for which the value of this reaction rate is $V_M/2$. The smaller the K_S, the greedier the enzyme for its substrate. We see from (1.14) that V_M is not an intrinsic constant of the enzyme, but is proportional to the initial quantity of free enzyme, E_o, whereas K_S is a constant characteristic of the enzyme, the so-called Michaelis constant. The net effect of adding an inhibitor to a solution of enzyme and substrate is to increase the apparent Michaelis constant, which becomes

$$K_S(1 + \frac{[I]}{K_I}) .$$

Up to now we have described the monoenzyme irreversible reaction (1.2), whose velocity term is given by (1.6), eventually inhibited by a competitive inhibitor, in which case the velocity term is given by (1.13). In fact the rate expressions for enzyme reactions in a well-stirred solution vary greatly. For example we shall deal with substrate inhibited phenomena for which the rate expression is given by:

$$(1.15) \qquad r = -\frac{d[S]}{dt} = V_M \frac{[S]}{K_S + [S](1 + \frac{[S]}{K_{SS}})}$$

Here K_{SS} is the inhibition constant of S for the enzyme.

More than one substrate can be involved in the reaction. A second substrate may be called a cosubstrate. Similarly, more than one product can result from the reaction. For example, with S = uric acid, A = oxygen, E = uricase, P = allantoin, we can write:

$$(1.16) \qquad S + A \xrightarrow{E} P + \text{other products.}$$

In the case of the uricase reaction (1.16), the rate expression may be taken as:

$$(1.17) \qquad r = -\frac{d[S]}{dt} = -\frac{d[A]}{dt} = V_M \frac{[A]}{K_A + [A]} \frac{[S]}{K_S + [S](1 + \frac{[S]}{K_{SS}})}$$

Here [A] is cosubstrate concentration and K_A is the Michaelis constant of enzyme E for cosubstrate A.

If [A] is small with respect to K_A, we can approximate (1.17) by:

$$(1.18) \qquad r = V_M \frac{[A]}{K_A} \frac{[S]}{K_S + [S](1 + \frac{[S]}{K_{SS}})}$$

If A is in excess, i.e. if [A] is large with respect to K_A, (1.17) reduces to (1.15). In the following we shall choose K_S as the unit of concentration, and use the dimensionless quantities:

(1.19) $s = [S]/K_S, \quad i = [I]/K_S, \quad a = [A]/K_S.$

The rates of reaction (1.6), (1.13), (1.15), (1.17), and (1.18) can then be written:

$$(1.20) \quad r = V_M \frac{s}{1 + s},$$

$$(1.21) \quad r = V_M \frac{s}{1 + \frac{K_S}{K_I} i + s},$$

$$(1.22) \quad r = V_M \frac{s}{1 + s + ks^2} \quad (k = \frac{K_S}{K_{SS}})$$

$$(1.23) \quad r = V_M \frac{a}{\frac{K_A}{K_S} + a} \frac{s}{1 + s + ks^2}$$

$$(1.24) \quad r = V_M \frac{K_S}{K_A} a \frac{s}{1 + s + ks^2}.$$

For more details on enzyme models, see [7], [8], and, mainly [9].

1.3 Enzyme Membranes

Enzyme kinetics described in the preceding section have been obtained by studying enzyme reactions in well-stirred solutions of the species involved (enzymes, substrates, products, inhibitors, activators). However, as pointed out in Section 1.1, in living organisms enzyme reaction is coupled to diffusion. Thus, after the first and necessary step of studying enzyme kinetics in an homogeneous phase, there is a need for a second step which would be enzyme kinetics in an heterogeneous phase [5]. In other terms, after studying "lumped" systems governed by ordinary differential equations, we now have to study "distributed" systems governed by partial differential equations.

Our tool will be artificial enzyme membranes, more precisely membranes in which enzymes are insolubly embedded by one of several means [10]. In particular the binding of enzyme molecules to inactive protein molecules under the action of glutaraldehyde produces a membrane [11].

Without entering into details, we can say the following about the recipe for preparing a membrane [11]: you make an homogeneous solution of inactive protein (albumin for example), enzyme (glucose oxidase for example, and glutaraldehyde, then, as for a pancake, you pour the mixture on a flat surface (glass). You wait several hours, until the water has evaporated, and obtain a membrane where glutaraldehyde acts as a glue to bind the enzyme and albumin molecules. This membrane presents the aspect of a translucid film of thickness $L = 50\mu$, with good mechanical properties. Figure 1.2 shows an electron microscope image of such a membrane. Remark the regular aspect and the absence of pores. Within this membrane, enzyme molecules are uniformly embedded. Due to their proteic environment, they exhibit an increased stability and remain active for a longer time than in solution. Many enzymes can be immobilized ([10], [12]).

1.4 Simple Irreversible Monoenzymatic Reactions

For the simple biochemical model to be described in this section, we shall assume that an irreversible monoenzymatic reaction, such as (1.2), takes place. For example glucose oxidase (E) catalyzes the transformation of glucose (S) in gluconic acid (P). The rate expression, in solution, is given by (1.6).

The membrane M separates two compartments, I and II, as depicted in Figure 1.3. Compartments I and II are 5 to 10 cm. long and several centimeters in height. The compartments contain well-stirred solutions of S, of concentrations S_1 and S_2 respectively. The membrane being initially empty of any substrate or product, the substrate moves in and is altered under the catalytic action of the enzyme. The membrane shape, with its slab geometry, is particularly convenient for an analysis of the phenomenon, inasmuch as it favors the direction transverse to the membrane. Letting $S(x,t)$ denote the concentration of substrate at (x,t) in this one-dimensional membrane, we have [13]:

$$(1.25) \qquad \frac{\partial S}{\partial t} = \left(\frac{\partial S}{\partial t}\right)_{\text{diffusion}} + \left(\frac{\partial S}{\partial t}\right)_{\text{reaction}} .$$

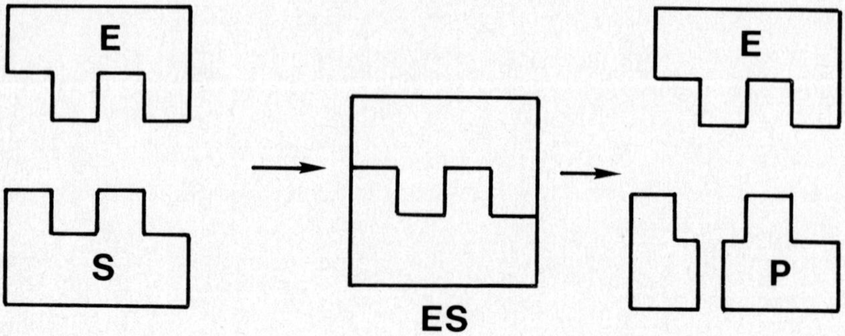

Fig. 1.1

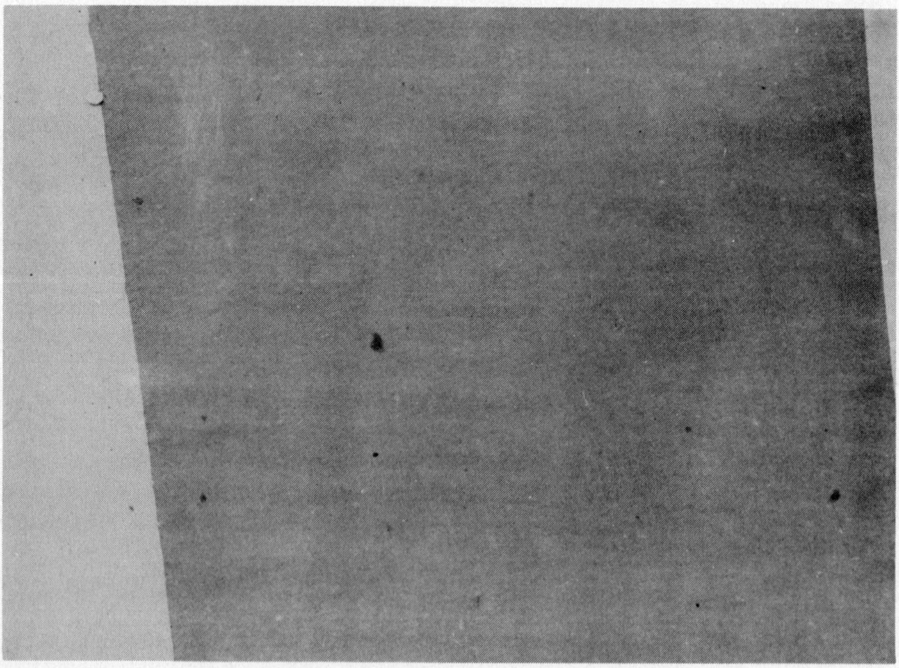

Fig. 1.2

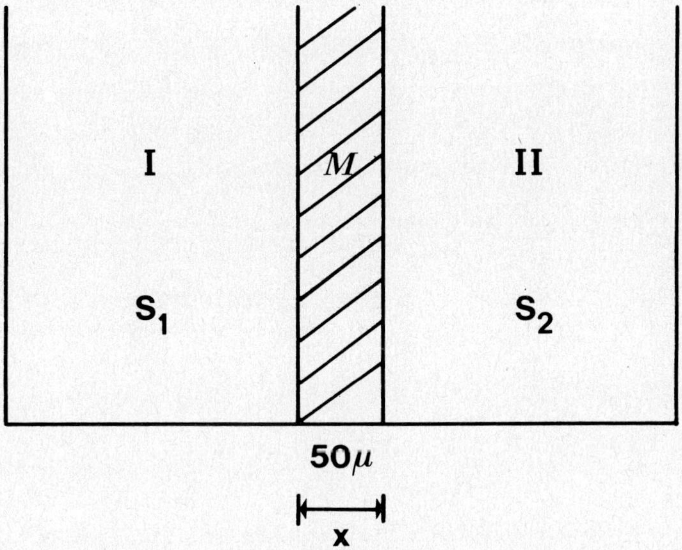

Fig. 1.3

We can assume that the coefficient of diffusion D_S is constant since the medium is homogeneous. Using Fick's law, we have:

$$(1.26)\qquad \left(\frac{\partial S}{\partial t}\right)_{\text{diffusion}} = D_S \frac{\partial^2 S}{\partial x^2} .$$

The velocity due to reaction is given by:

$$(1.27)\qquad \left(\frac{\partial S}{\partial t}\right)_{\text{reaction}} = -V_M\, S/(K_M + S)$$

Thus the evolution of S in the membrane is given by:

$$(1.28)\qquad \frac{\partial S}{\partial t} = D_S \frac{\partial^2 S}{\partial x^2} - V_M \frac{S}{K_S + S}, \qquad 0 < x < L, \quad t > 0.$$

Here L is the membrane thickness.

If we introduce the non-dimensional variables:

$$(1.29)\qquad s = S/K_S,\ x' = x/L,\ \ t' = t/(L^2/D_S),$$

substitute them into (1.28) and for convenience drop the primes, we obtain the dimensionless equation:

$$(1.30) \qquad \frac{\partial s}{\partial t} - \frac{\partial^2 s}{\partial x^2} + \sigma \frac{s}{1 + s} = 0, \qquad 0 < x < 1, \quad t > 0,$$

where the non-dimensional constant σ is defined by:

$$(1.31) \qquad \sigma = \frac{V_M}{K_S} \frac{L^2}{D_S} .$$

σ, square of the so-called Thiele modulus, is the ratio of L^2/D_S, characteristic time for diffusion, and K_S/V_M, characteristic time for reaction. One may ask whether it is realistic to compare the coupling of reaction and diffusion within cell membranes, which are 75 to 100 x $10^{-4}\mu$ in thickness, and within artificial membranes of thickness 50μ. In fact the single parameter σ is of importance for the behavior of the membrane. For an artificial enzyme membrane, any desired value of σ may be achieved by an appropriate choice of V_M, which is at our disposal since it is proportional to the concentration E_o of enzyme added to inactive protein and glutaraldehyde when preparing the membrane.

For the simple biochemical model we have in mind, namely a glucose oxidase membrane, the parameter values are $L = 5\ 10^{-3}$cm, $D_S = 5\ 10^{-3}$ cm^2h^{-1}(h = hour), $V_M = 3.66\ 10^{-3}$moles $cm^{-3}h^{-1}$, and $K_M = 1.3\ 10^{-5}$ moles cm^{-3}, whence $\sigma = 1.4$. The characteristic time for diffusion $L^2/D_S = 5\ 10^{-3}$ h = 18 seconds.

Boundary conditions: if one assumes that the concentrations S_1 and S_2 are held fixed in compartments I and II, one obtains "bath" or Dirichlet conditions:

$$(1.32) \qquad s(0,t) = \alpha, \ s(1,t) = \beta, \qquad t > 0,$$

where: $\alpha = S_1/K_S$ and $\beta = S_2/K_S$. Such boundary conditions are also appropriate if solution volumes in the compartments are large compared with membrane volume and if the experiment is of short duration. A configuration of interest is a membrane immersed in a solution of substrate. In this case both edges of the membrane are exposed to the same concentration α, concentration of substrate in the bulk solution, and we still have (1.32), with $\beta = \alpha$.

Initial conditions: the membrane being initially empty of any substrate, the initial condition is:

$$(1.33) \qquad s(x,0) = 0, \qquad\qquad 0 < x < 1.$$

Product concentration: one can easily see that the product P concentration (again in dimensionless form) is given by:

$$(1.34) \quad \begin{cases} \dfrac{\partial p}{\partial t} - \dfrac{D_P}{D_S}\dfrac{\partial^2 p}{\partial x^2} = \sigma \dfrac{s}{1+s}, & 0 < x < 1, \quad t > 0 \\ p(0,t) = p(1,t) - 0 & \\ p(x,0) = 0. & \end{cases}$$

Here $p = [P]/K_S$, D_p is the diffusion coefficient of P and it is assumed that the membrane is initially empty of any product, and the adjacent compartments are large enough with respect to membrane volume for P concentration to remain 0 within them, although P may flow across the membrane edges.

Thus S and P concentrations in an artificial membrane separating two compartments, and where a simple irreversible monoenzymatic reaction takes place, are governed by equations (1.30), (1.32), (1.33),(1.34). We mentioned that it was also the case for a membrane immersed in a solution of substrate.

Enzyme electrodes: Another configuration, important in many applications, also falls within this example: artificial membranes coupled with electrodes ([10], [14], [15], [16]). A membrane is attached to an electrode which is sensitive to the product P of the reaction. In Figure 1.4 for example, the substrate (urea) is transformed by the enzymatic layer (urease and protein) which covers the electrode bulb: as a result, monovalent cations appear, which are detected by the electrode. When the electrode is in contact with a solution containing the substrate to be measured, S and P concentrations within the membrane are governed by equations:

$$(1.37)\quad \begin{cases} s_t - s_{xx} + \sigma F(s) = 0, \quad p_t - p_{xx} - \sigma F(s) = 0, \\ s(0,t) = \alpha, \quad p(0,t) = 0, \quad s_x(1,t) = p_x(1,t) = 0, \\ s(x,0) = p(x,0), \quad 0 < x < 1, \quad t > 0, \\ F(s) = s/(1 + |s|). \end{cases}$$

Here and hereafter, $s_t = \frac{\partial s}{\partial t}$, $s_x = \frac{\partial s}{\partial x}$, $s_{xx} = \frac{\partial^2 s}{\partial x^2}$. Due to the presence of the electrode, there is a zero-flux boundary condition at the impermeable wall $x = 1$. The S and P profiles evolve towards a stable steady-state, as depicted in Figure 1.5, where substrate (---) and product (——) concentration profiles, solutions of (1.37) for $\sigma = 0.35$ and $\alpha = \beta = 1$, are represented at times $t = 0.1$, 0.2, and 0.7. At time $t = 0.7$ the profiles have nearly attained the steady state. If P is a cation to which the electrode is sensitive, from the measurement of the electrode at steady state, one can get the concentration of P against the electrode and hence of S in bulk solution. Other electrodes, pO_2 electrodes, are sensitive to the oxygen pressure. Glucose electrodes can be made with such an electrode coupled with a membrane containing glucose oxidase. The reaction:

$$(1.38)\quad \text{glucose} + O_2 \xrightarrow{\text{glucose oxidase}} \text{gluconic acid} + H_2O_2$$

consumes oxygen, so that the oxygen pressure is less at the contact of the electrode ($x = 1$) than in solution ($x = 0$). A well-defined relation exists between the pO_2 level, at $x = 1$ at steady state, and glucose concentration in the bulk solution. Due to the specificity of the enzyme, such an electrode is sensitive to glucose only. It is thus possible to prepare electrodes sensitive to one particular substrate (glucose, maltose, saccharose, lactose, etc.) and to continuously monitor the concentration of these substances in media like blood or the broth of fermentations.

Later is this chapter we shall be interested in the numerical and mathematical analysis of the systems introduced in this section, more precisely the membrane modeled by equations (1.30), (1.32), (1.33), and (1.34), and the electrode modeled by equations (1.37). In each case p is defined by a linear parabolic equation. Thus we shall restrict ourselves to the equations giving s, the definition of p being straightforward once s is known. These equations are (1.1) and its

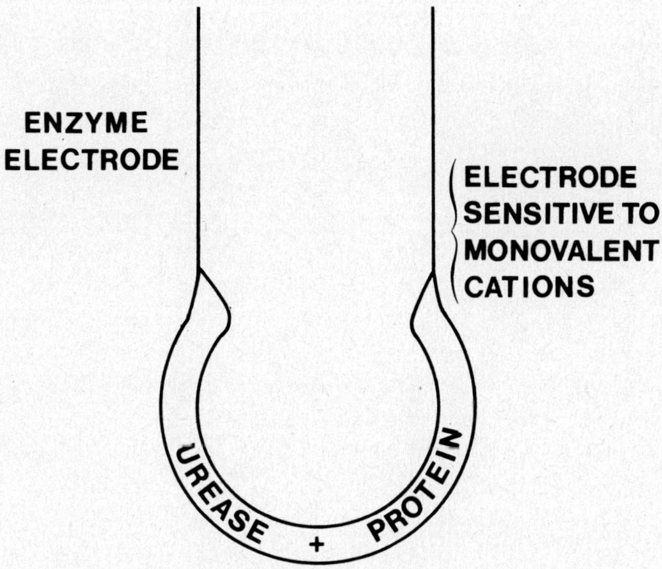
ENZYME
ELECTRODE
ELECTRODE
SENSITIVE TO
MONOVALENT
CATIONS
UREASE
+
PROTEIN

Fig. 1.4

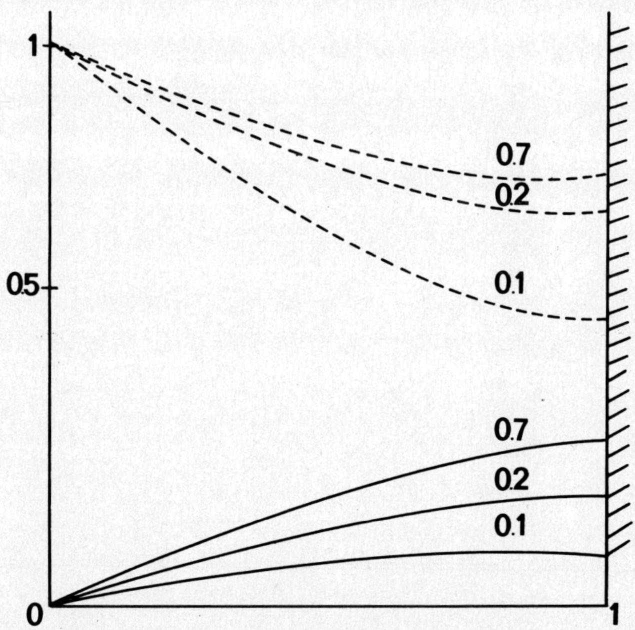
1
0.5
0
1
0.7
0.2
0.1
0.7
0.2
0.1

Fig. 1.5

variant with a 0-flux boundary condition at $x = 1$.

1.5 Numerical Solution

Because the concentration profiles within the membranes are not directly observable, biochemists are very interested by the numerical simulation of equations (1.1). Indeed the only quantities that biochemists can measure are $-s_x(0,t)$ and $s_x(1,t)$, fluxes of substrate entering into the membrane from the compartments, or, in the case of an electrode, S or P concentration along the electrode, namely $s(1,t)$ or $p(1,t)$. Fortunately, there exist simple and efficient numerical methods for solving equations like (1.1), the so-called explicit and implicit finite difference methods [17]. Although they are well-known, we describe them here for the sake of completeness. They are very useful for solving, not only (1.1), but also similar reaction diffusion problems. In order to show how simple they are we give, in Figures 1.6 and 1.7, listings of the corresponding Fortran programs.

1.5.1 - Explicit scheme: We divide the interval $[0,1]$ into N equal intervals of length $h = \Delta x = 1/N$, and define a time step $k = \Delta t$. We call s_i^n an approximation of $s(ih,nk)$. This approximation is defined by the following explicit scheme:

$$(1.39)\quad \begin{cases} \dfrac{s_i^{n+1} - s_i^n}{k} - \dfrac{s_{i+1}^n + s_{i-1}^n - 2s_i^n}{h^2} + \sigma \dfrac{s_i^n}{1 + |s_i^n|} = 0, & 1 \leq i \leq N-1,\ n \geq 0 \\ s_o^n = \alpha,\ s_N^n = \beta, \quad n \geq 0 & \\ s_i^o = 0, & 1 \leq i \leq N-1. \end{cases}$$

It is easily seen that s_i^{n+1} is explicitly known once the three values s_{i-1}^n, s_i^n, and s_{i+1}^n are known. As $s_o^o = \alpha$, $s_i^o = 0$, $1 \leq i \leq N-1$, and $s_N^o = \beta$, once can calculate the s_i^1, $1 \leq i \leq N-1$, and so on: once the s_i^n are known at time level n, the s_i^{n+1} may be calculated at time level $n + 1$ (Figure 1.6), by formula:

$$(1.40)\quad s_i^{n+1} = \left(1 - 2r - \frac{\sigma k}{1 + |s_i^n|}\right) s_i^n + r(s_{i-1}^n + s_{i+1}^n),\ r = k/h^2.$$

```
      DIMENSION S(21)
      F(C)=SIGMA*C/(1.+C)
      NX=20
      NT=1000
      ITP=100
      DT=0.001
      SIGMA=1.4
      S(1)=1.
      S(NX+1)=1.
      DO 50 I=2,NX
50    S(I)=0.
      R=DT*(FLOAT(NX))**2
      DO 200 IT=1,NT
      GRAD2=S(2)-S(1)
      DO 100 I=2,NX
      GRAD1=GRAD2
      GRAD2=S(I+1)-S(I)
100   S(I)=S(I)+R*(GRAD2-GRAD1)-DT*F(S(I))
C     IN CASE OF 0-FLUX B.C. AT X=1,INSERT HERE
C     S(NX+1)=S(NX+1)-2.*R*GRAD1-DT*F(S(NX+1))
      IF (MOD(IT,ITP).NE.0) GOTO 200
      WRITE (6,1000) IT
      WRITE (6,1100) S
200   CONTINUE
1000  FORMAT(/' TIME STEP #',I4/)
1100  FORMAT(1X,11E10.3)
      STOP
      END
```

Fig. 1.6

We may remark that if the s_i^n are positive, so are the s_i^{n+1}, provided:

$$k < h^2/(2 + \sigma h^2). \tag{1.41}$$

This stability condition ensures a good behavior of the explicit scheme. (In the case $\sigma = 0$ it reduces to the well known stability condition $k/h^2 < 1/2$). For example, if $N = 20$, $h = 0.05$, $\sigma = 1.4$, the stability condition is $k < 0.0013$. We choose $k = 0.001$.

This drawback of being constrained to choose small time increments may be overcome by using an implicit scheme.

1.5.2 - Implicit schemes: There are many of them. Here we give the totally implicit finite difference scheme:

(1.42)
$$\begin{cases} \dfrac{s_i^{n+1} - s_i^n}{k} - \dfrac{s_{i+1}^{n+1} + s_{i-1}^{n+1} - 2s_i^{n+1}}{h^2} + \sigma \dfrac{s_i^{n+1}}{1 + |s_i^{n+1}|} = 0, \quad 1 \le i \le N-1, \quad n \ge 0, \\ s_o^{n+1} = \alpha, \quad s_N^{n+1} = \beta, \quad n \ge 0 \\ s_i^o = 0, \quad 1 \le i \le N-1. \end{cases}$$

Here the s_i^{n+1} are no more explicitly defined but, in order to find them at each time level n+1, we have to solve the set of N-1 nonlinear equations:

(1.43)
$$\begin{cases} \dfrac{u_i - s_i^n}{k} - \dfrac{u_{i+1} + u_{i-1} - 2u_i}{h^2} + \sigma \dfrac{u_i}{1 + |u_i|} = 0, \quad 1 \le i \le N-1, \\ u_o = \alpha, \quad u_N = \beta. \end{cases}$$

A possible iterative method for solving (1.43) is the Newton method. Easier to implement is the following one:

i/ take as starting values $u_i^o = s_i^n$, $1 \le i \le N-1$.

ii/ once the u_i^m are known, define the u_i^{m+1} as the solution of the <u>linear</u> system:

(1.44)
$$\begin{cases} \dfrac{u_i^{m+1} - s_i^n}{k} - \dfrac{u_{i+1}^{m+1} + u_{i+1}^{m+1} - 2u_i^{m+1}}{h^2} + \sigma \dfrac{u_i^{m+1}}{1 + |u_i^m|} = 0, \quad 1 \le i \le N-1, \\ u_o^{m+1} = \alpha, \quad u_N^{m+1} = \beta. \end{cases}$$

iii/ Stop when $\max_{1 \le i \le N-1} |u_i^{m+1} - u_i^m| < \varepsilon$, ε being some error tolerance.

In fact, in the program shown in Figure 1.7, it was decided to adopt $s_i^{n+1} = u_i^1$, $1 \le i \le N-1$, $n \ge 0$.

Some words about the solution of the linear system (1.44). It takes only 7 lines of Fortran (lines 17-23 in Figure 1.7) to efficiently solve this tridiagonal system by Gauss elimination. Let us briefly recall the method [17] for the tridiagonal system:

$$(1.45)\quad \begin{cases} -a_i u_{i-1} + b_i u_i - c_i u_{i+1} = d_i, & 1 \leq i \leq N-1 \\ u_o = \alpha, \quad u_N = \beta & \end{cases}$$

```
 1              DIMENSION S(21),E(20),F(20)
 2              NX=20
 3              NT=1000
 4              ITP=100
 5              DT=0.001
 6              SIGMA=1.4
 7              S(1)=1.
 8              S(NX+1)=1.
 9              DO 50 I=2,NX
10    50        S(I)=0.
11              R=DT*(FLOAT(NX))**2
12              C1=1.+2.*R
13              C2=SIGMA*DT
14              E(1)=0.
15              F(1)=1.
16              DO 200 IT=1,NT
17              DO 100 I=2,NX
18              DEN=C1+C2/(1.+S(I))-R*E(I-1)
19              E(I)=R/DEN
20    100       F(I)=(S(I)+R*F(I-1))/DEN
21              DO 150 J=2,NX
22              I=NX+2-J
23    150       S(I)=E(I)*S(I+1)+F(I)
24              IF (MOD(IT,ITP).NE.0) GOTO 200
25              WRITE (6,1000) IT
26              WRITE (6,1100) S
27    200       CONTINUE
28    1000      FORMAT(/' TIME STEP #',I4/)
29    1100      FORMAT(1X,11E10.3)
30              STOP
31              END
```

Fig. 1.7

Equations (1.45) may be rewritten:

$$(1.46)\quad u_i = E_i u_{i+1} + F_i, \qquad 0 \leq i \leq N-1$$

This is evident for i = 0, with:

$$(1.47)\quad E_o = 0 \text{ and } F_o = \alpha$$

This is true for i if it is also true for i-1. Indeed, if $u_{i-1} =$

$E_{i-1}u_i + F_{i-1}$, then plugging this value of u_{i-1} into (1.45), we obtain:

$$-a_i(E_{i-1}u_i + F_{i-1}) + b_iu_i - c_iu_{i+1} = d_i,$$

whence (1.46) with:

(1.48) $$E_i = \frac{c_i}{b_i - a_iE_{i-1}} \quad \text{and} \quad F_i = \frac{d_i + a_iF_{i-1}}{b_i - a_iE_{i-1}}.$$

Thus (1.46) is proven by recurrence and the algorithm for solving (1.45) is the following:

i/ using formulas (1.47) and (1.48), calculate the E_i, F_i, $1 \leq i \leq N-1$ (triangularization),

ii/ using formula (1.46), and knowing u_N, calculate u_{N-1}, then $u_{N-2}, \ldots, u_1$ (backward substitution).

Although this algorithm is slightly more complex than the former (31 lines of program in Figure 1.7 instead of 28 lines in Figure 1.6!) and more computer time consuming at each level of time (there may be several iterations with respect to m and within each of these iterations a certain amount of operations are required), however the overall advantage comes from the fact that we are no more constrained to small time steps k, but may choose any value for this increment of time. For example we may choose k = 0.01, which is forbidden in the explicit method.

Of course with k very large we do not find the steady state solution in one step, but have to iterate several times. The successive profiles thus obtained do not have much to do with the transient evolution of the system, except that they represent closer and closer approximations of the steady state solution in what is in fact a fixed point algorithm to calculate this steady state solution.

Remark 1.1 - No-flux boundary condition $s_x(1,t) = 0$: A slight modification of the numerical schemes described in (1.39) and (1.42) enables handling no-flux boundary conditions. In both cases we only have to write equation (1.39) (resp. (1.42)) for i = N with $s^n_{N+1} = s^n_{N-1}$ (resp. $s^{n+1}_{N+1} = s^{n+1}_{N-1}$). In the last case trivial modifications have to be made. Since the last equation is:

$$-a_N u_{N-1} + b_N u_N - c_N u_{N-1} = d_N$$

and we have

$$u_{N-1} = E_{N-1} u_N + F_{N-1} ,$$

we obtain:

$$(1.49) \qquad u_N = \frac{d_N + (a_N + c_N) F_{N-1}}{b_N - (a_N + c_N) E_{N-1}} .$$

1.5.3 - Numerical results: Application of either method provides, at each level of time, a profile of concentration as depicted in Figures 1.5 and 1.8. Figure 1.8 shows the evolution of the substrate concentration profile in an enzyme membrane shortly after introduction into a substrate solution. Here $\sigma = 1.4$, and $\alpha = \beta = 1$. The time increment is $\Delta t = 0.001$ and at time $700\ \Delta t$ $(= 0.7)$ the concentration profile is nearly stationary. The existence of such concentration profiles can be shown by electron microscope images of the membrane (Figure 1.9) ([18], [19], [20]). For other electron microscopy studies, see [21] and [22]. Figure 1.9 shows an electron microscope image of a membrane exhibiting a concentration profile: the membrane boundaries are highly contrasted and the density of the black points (due to insoluble product) is much lower in the middle.

We may remark that the profiles found for the electrode system (1.37) for a given value of σ are the same as the left-hand half parts of the profiles found for the system (1.1) with $\beta = \alpha$ and σ four times larger.

1.6 Mathematical Methods

For reaction-diffusion problems, there exist well-known methods for studying existence and uniqueness of a positive solution, and asymptotic stability of a steady state [23]. Here we shall present methods based upon the existence of upper and lower solutions, as explained by Sattinger in [24] and [25], Amann in [26], Fife in [6], and Pao in [27], although existence and uniqueness of a solution for problem (1.1) results classically from the monotonicity of the function:

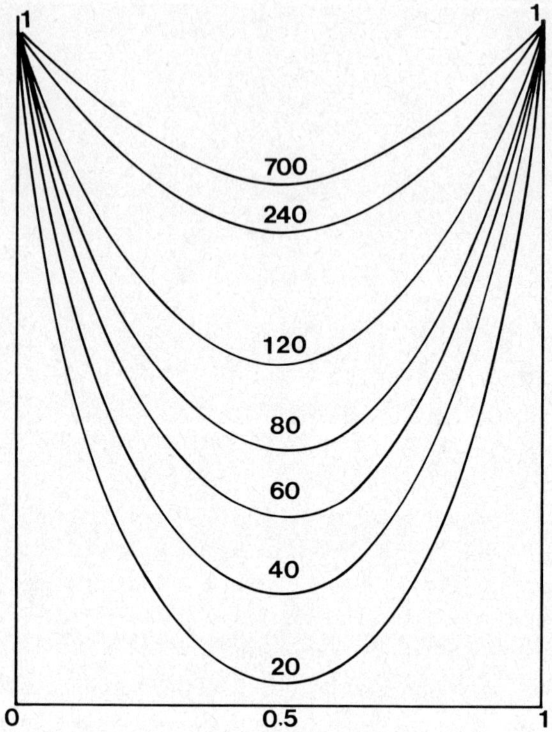

Fig. 1.8

Fig. 1.9

$s \to s/(1 + |s|)$ [28]. In fact the notions and results introduced in this section, particularly theorem 1.1, will be useful for subsequent problems, more complex than the "model" case (1.1):

$$(1.50)\quad \begin{cases} s_t(x,t) - s_{xx}(x,t) + \sigma F(s(x,t)) = 0, & 0 < x < 1, \quad t > 0, \\ s(0,t) = \alpha, \quad s(1,t) = \beta, & t > 0, \\ s(x,0) = 0, & 0 < x < 1, \\ F(s) = s/(1 + |s|), \quad \alpha > 0, \ \beta > 0, & \sigma > 0. \end{cases}$$

We shall show that the evolution problem (1.50) admits a positive solution and only one, a property we must expect because of the physical origin of this problem, and that the transient concentration profile tends, as $t \to \infty$, towards a stationary profile, which is an unique solution of:

$$(1.51)\quad \begin{cases} -s''(x) + \sigma F(s(x)) = 0, & 0 < x < 1 \\ s(0) = \alpha, \quad s(1) = \beta. \end{cases}$$

We shall essentially rely upon the notions of upper and lower solutions for problems (1.50) and (1.51).

An upper solution for (1.50) is a function $\hat{v}(x,t)$ such that:

$$(1.52)\quad \begin{cases} \hat{v}_t - \hat{v}_{xx} + \sigma F(\hat{v}) \geq 0, & 0 < x < 1, \quad t > 0 \\ \hat{v}(0,t) \geq \alpha, \quad \hat{v}(1,t) \geq \beta, \\ \hat{v}(x,0) \geq 0, \end{cases}$$

whereas a lower solution satisfies the reversed inequalities.

Let us define:

$$(1.53)\quad \phi(x) = \alpha(1-x) + \beta x.$$

It is easily seen that $\hat{v}(x,t) = \phi(x)$ is an upper solution for the parabolic problem (1.50), while $\bar{v}(x,t) = 0$ is a lower solution. As will follow from Proposition 1.1, this is enough to claim that there is an unique solution $s(x,t)$ to (1.50), satisfying:

$$(1.54) \quad 0 \le s(x,t) \le \phi(x).$$

Similarly, upper and lower solutions to (1.51) are functions $\hat{v}(x)$ and $\bar{v}(x)$ satisfying respectively:

$$(1.55) \quad \begin{cases} -\hat{v}''(x) + \sigma F(\hat{v}(x)) \ge 0, & 0 < x < 1, \\ \hat{v}(0) \ge \alpha, \quad \hat{v}(1) \ge \beta. \end{cases}$$

and the reversed inequalities.

Again $\phi(x)$ and 0 are upper and lower solutions for (1.51), and, as will follow from Proposition 1.2, this is enough to ensure the existence of at least one solution $s(x)$ for (1.51) such that:

$$(1.56) \quad 0 \le s(x) \le \phi(x).$$

Uniqueness of this stationary solution will result from the monotonicity of function F and Proposition 1.3.

To begin with, let us state four propositions, which will be proven later in this chapter.

Proposition 1.1 - Let $\bar{y}$ and $\hat{y}$ be a pair of lower and upper solutions such that $\bar{y} \le \hat{y}$ for the evolution problem:

$$(1.57) \quad \begin{cases} s_t - s_{xx} + \sigma F(s) = 0, & 0 < x < 1, \quad t > 0, \\ s(0,t) = \alpha, \quad s(1,t) = \beta, \\ s(x,0) = s_o(x) & (s_o \in L^2(0,1)) \end{cases}$$

where F is Lipschitz-continuous. Then problem (1.57) has an unique solution $s(x,t)$ such that:

$$(1.58) \quad \bar{y}(x,t) \le s(x,t) \le \hat{y}(x,t), \quad 0 < x < 1, \quad t > 0.$$

Proposition 1.2 - Let $\bar{y}$ and $\hat{y}$ be a pair of lower and upper solutions such that $\bar{y} \le \hat{y}$ for the stationary problem:

$$(1.59) \quad \begin{cases} -s''(x) + \sigma F(s(x)) = 0, & 0 < x < 1, \\ s(0) = \alpha, \quad s(1) = \beta, \end{cases}$$

where F is Lipschitz-continuous. Then problem (1.59) has at least one solution s(x) such that:

(1.60) $\bar{y}(x) \leq s(x) \leq \hat{y}(x), \qquad 0 < x < 1.$

Proposition 1.3 - Moreover, if F is monotone increasing, then the solution of (1.59) is unique.

Remark 1.2 - Of course, in Proposition 1.1, we must have:

(1.61) $\bar{s}(x,0) \leq s_o(x) \leq \hat{s}(x,0).$

Remark 1.3 - F is said to be Lipschitz-continuous if there exists a constant L such that:

(1.62) $|F(s_1) - F(s_2)| \leq L|s_1 - s_2|$

for every $s_1, s_2 \in R$.

Without loss of generality we can assume that L = 1, by choosing σ conveniently.

Remark 1.4 - F is said to be monotone increasing if:

(1.63) $s_1 < s_2 \Longrightarrow F(s_1) \leq F(s_2)$

The way of applying Proposition 1.1 to problem (1.50) is quite apparent. The Lipschitz continuity of F is a consequence of $|F'(s)| \leq 1$, whence $|F(s_1) - F(s_2)| \leq |s_1 - s_2|$. Similarly Propositions 1.2 and 1.3 apply obviously to problem (1.51).

The relation between the transient profile s(•,t), solution of (1.50), and the steady-state profile s_e, solution of (1.51), is given by the following proposition:

Proposition 1.4 - Let s_e be the solution of (1.51). Then the solution s of (1.50) satisfies the relation:

$$s_e(x) - re^{-\mu t}w(x) \leq s(x,t) \leq s_e(x) + re^{-\mu t}w(x), \quad 0 < x < 1,$$

for t large enough, where r is a non-negative constant and w is a positive eigenfunction corresponding to the least eigenvalue of the eigenvalue problem:

$$-w''(x) = \mu w(x), \qquad 0 < x < 1, \quad w(0) = w(1) = 0.$$

Remark 1.5 - Similar results hold when we have a no-flux boundary condition at $x = 1$. In this case we take, as an upper solution, the constant function $\hat{y}(x,t) \equiv \alpha$.

The remainder of this chapter is concerned with the statement and proof of Theorem 1.1, a most useful theorem due to Amann [26], and the derivation of Propositions 1.1 - 1.4 from this theorem.

Theorem 1.1 - Let (E,P) be an O.B.S. and let $[\bar{y},\hat{y}]$ be a nonempty order interval in E. Suppose that $f : [\bar{y},\hat{y}] \to E$ is an increasing compact map such that $\bar{y} \leq f(\bar{y})$ and $f(\hat{y}) \leq \hat{y}$. Then f possesses a minimal fixed point $\bar{x}$ and a maximal fixed point $\hat{x}$. Moreover,

$$\bar{x} = \lim_{k\to\infty} f^k(\bar{y}) \quad \text{and} \quad \hat{x} = \lim_{k\to\infty} f^k(\hat{y}),$$

and the sequence $(f^k(\bar{y}))$ is increasing and $f^k(\hat{y}))$ is decreasing.

Comments:

Let us first explain the terms which occur in this statement. E is a real Banach space. A subset P of E is a cone if:

$$P + P \subset P, \quad R_+P \subset P, \quad P \cap (-P) = \{0\}, \quad \bar{P} = P.$$

P induces an ordering in E, defined by: $x \leq y \Leftrightarrow y - x \in P$. Endowed with this ordering, E is called an ordered Banach space (O.B.S.), denoted (E,P).

The order interval $[\bar{y},\hat{y}]$ is the set of those z such that $\bar{y} \leq x \leq \hat{y}$. A map f of the O.B.S. (E,P) into the O.B.S. (F,Q) is said to be increasing if $x \leq y \Rightarrow f(x) \leq f(y)$.

X being a subset of E, f is a compact map of X into F if f is continuous and $\overline{f(X)}$ is compact.

$\bar{x}$ and $\hat{x}$ are called minimal and maximal fixed points of f in $[\bar{y},\hat{y}]$ if every fixed point y of f in $[\bar{y},\hat{y}]$ satisfies $\bar{x} \leq y \leq \hat{x}$.

Proof of Theorem 1.1 - From the sequence $(f^k(\bar{y}))$ we can extract a subsequence converging towards $\bar{x}$, $\bar{x} \in E$. As $\bar{y} \leq f^k(\bar{y}) \leq \hat{y}$, and as the cone P defining the ordering is closed, $\bar{y} \leq \bar{x} \leq \hat{y}$. It is immediate that the whole sequence is converging towards $\bar{x}$. Similarly the whole sequence $(f^k(\hat{y}))$ converges towards an element $\hat{x}$ in $[\bar{y},\hat{y}]$ which, since $f^k(\bar{y}) \leq f^k(\hat{y})$, satisfies $\bar{x} \leq \hat{x}$. The fact that $\bar{x}$ is minimal follows from the consideration of a candidate fixed point x in $[\bar{y},\hat{y}]$. Since $x = f(x)$, we can consider x as satisfying $f(x) \leq x$, and make the same analysis as above with the interval $[\bar{y},x]$ instead of $[\bar{y},\hat{y}]$. We shall find the same $\bar{x}$, so that $\bar{x} \leq x$. Similarly $x \leq \hat{x}$.

Applications of Theorem 1.1:

Proof of Proposition 1.1 - We are going to prove the existence and uniqueness of a positive solution for (1.57) on the time interval $]0,T[$, T arbitrary, whence the property on the time interval $]0,+\infty[$. We take $E = L^2(Q)$, where $Q =]0,1[\times]0,T[$, $T > 0$. $P = \{v \in L^2(Q) \mid v(x,t) \geq 0 \text{ a.e.}\}$, $f : u \to v$ defined by:

$$(1.64) \quad \begin{cases} v_t - v_{xx} + \sigma v = \sigma u - \sigma F(u), \\ v(0,t) = \alpha, \quad v(1,t) = \beta, \\ v(x,0) = s_o(x) \qquad \text{(we suppose } s_o \in L^2(0,1)\text{)}. \end{cases}$$

Let us check that the hypotheses of Theorem 1.1 are satisfied.

i/ f is increasing: suppose $u_1 \leq u_2$ and define $v_i = f(u_i)$, $i = 1,2$ by equations (1.64). The difference $w = v_2 - v_1$ satisfies:

$$\begin{cases} w_t - w_{xx} + \sigma w = \sigma(u_2 - u_1) - \sigma(F(u_2) - F(u_1)) \geq 0 \\ w(0,t) = w(1,t) = 0, \\ w(x,0) = 0 \end{cases}$$

where the inequality results from the Lipschitz continuity of F. The maximum principle for parabolic problems implies that $w \geq 0$.

ii/ f is compact: the map $u \to v - \phi$ is continuous from $L^2(Q)$ to

$W(0,T)$, where $\phi = \alpha(1-x) + \beta x$, $W(0,T) = \{w | w \in L^2(0,T;V),$ $w_t \in L^2(0,T;V')\}$, $V = H_0^1(\Omega)$, $V' = H^{-1}(\Omega)$, $\Omega =]0,1[$, and the injection from $W(0,T)$ to $L^2(Q)$ is compact [28]. Thus the map of $f : u \to v$ from E to E is compact.

iii/ $\underline{\bar{y} \le f(\bar{y})}$ and $\underline{f(\hat{y}) \le \hat{y}}$: in fact these inequalities are equivalent to the properties for $\bar{y}$ and $\hat{y}$ to be lower and upper solutions. Let us show for example that if $\hat{y}$ is an upper solution for problem (1.57), then $f(\hat{y}) \le \hat{y}$. Let $v = f(\hat{y})$.

We have:

$$(1.65)\quad \begin{cases} v_t - v_{xx} + \sigma v = \sigma\hat{y} - \sigma F(\hat{y}), \\ v(0,t) = \alpha, \quad v(1,t) = \beta, \\ v(x,0) = s_o(x) \end{cases}$$

and

$$(1.66)\quad \begin{cases} \hat{y}_t - \hat{y}_{xx} + \sigma F(\hat{y}) \ge 0, \\ \hat{y}(0,t) \ge \alpha, \quad \hat{y}(1,t) \ge \beta, \\ \hat{y}(x,0) \ge s_o(x). \end{cases}$$

Subtracting (1.65) from (1.66) gives, for $w = \hat{y} - v$,

$$(1.67)\quad \begin{cases} w_t - w_{xx} + \sigma w \ge 0, \\ w(0,t) \ge 0, \quad w(1,t) \ge 0, \\ w(x,0) \ge 0, \end{cases}$$

whence, from the maximum principle for parabolic equations, $w \ge 0$, or $\hat{y} - v \ge 0$, or, $f(\hat{y}) \le \hat{y}$.

Thus existence of at least one solution for (1.57) on $[0,T]$ results from application of Theorem 1.1. In order to prove uniqueness on the same time interval $[0,T]$, we again use the Lipschitz continuity of F. Let s_1 and s_2 be two solutions, and $w = s_1 - s_2$. Then:

$$(1.68)\quad \begin{cases} w_t - w_{xx} = \sigma(F(s_2) - F(s_1)), \\ w(0,t) = w(1,t) = 0, \\ w(x,0) = 0. \end{cases}$$

Hence:

$$\int_0^1 (w_t - w_{xx})w\,dx = \frac{1}{2}\frac{d}{dt}\left(\int_0^1 w^2 dx\right) + \int_0^1 w_x^2 dx =$$

$$\sigma \int_0^1 (F(s_2) - F(s_1))w\,dx \leq \sigma \int_0^1 w^2 dx.$$

Let: $\phi(t) = \int_0^1 w^2(x,t)dx.$

We have successively:

$$\phi'(t) \leq 2\sigma\phi(t), \quad e^{-2\sigma t}(\phi'(t) - 2\sigma\phi(t)) \leq 0,$$

$(e^{-2\sigma t}\phi(t))' \leq 0$ and the function $t \to e^{-2\sigma t}\phi(t)$ is decreasing.

But: $\phi(t) \geq 0$ and $\phi(0) = 0$, therefore $\phi(t) \equiv 0$.

Remark 1.6 - Thus the solution s of problem (1.57) belongs to $\phi + W(0,t)$. The only hypothesis on s_o is that $s_o \in L^2(0,1)$. In particular we do not need to have $s_o(0) = \alpha$ or $s_o(1) = \beta$. But on the other hand we know that as soon as $t > 0$, the profile $x \to s(x,t)$ is smooth in the sense that $s(\cdot,t) \in \phi + H_o^1(0,1) \subset C^0([0,1])$.

Proof of Proposition 1.2 - We take $E = L^2(\Omega)$, $\Omega =]0,1[$, $P = \{v \in E | v(x) > 0 \text{ a.e.}\}$, $\hat{y} = \alpha(1-x) + \beta x$, $\bar{y} = 0$, $f : u \to v$ defined by:

$$(1.69) \quad \begin{cases} -v'' + \sigma v = \sigma u - \sigma F(u), & 0 < x < 1, \\ v(0) = \alpha, \quad v(1) = \beta. \end{cases}$$

i/ f is increasing: suppose $u_1 \leq u_2$. We have:

$$-(v_2-v_1)'' + \sigma(v_2-v_1) = \sigma(u_2-u_1) - \sigma(F(u_2) - F(u_1)) \geq 0,$$

together with $v_2 - v_1 = 0$ at $x = 0$ and $x = 1$. Thus, from the maximum principle for elliptic problems [29], it follows that $v_2 - v_1 \geq 0$.

ii/ <u>f is continuous</u> from E to E since $u \to v$ is continuous from E to $H^1(\Omega)$ and $v \to v$ is continuous from $H^1(\Omega)$ to E.

iii/ <u>f is compact</u> from E to E since $u \to v$ is continuous from E to $H^1(\Omega)$ and $v \to v$ is compact from $H^1(\Omega)$ to E.

iv/ <u>$\bar{y} \leq f(\bar{y})$ and $f(\hat{y}) \leq \hat{y}$</u> result from the hypothesis that $\bar{y}$ and $\hat{y}$ are lower and upper solutions for problem (1.59), as in the proof of Proposition 1.1.

<u>Proof of Proposition 1.3</u> - Theorem 1.1 does not guarantee uniqueness of a fixed point for the map f defined by (1.69). However, if F is a monotone increasing function, we can prove that $\bar{s} = \hat{s}$, $\bar{s}$ and $\hat{s}$ being the minimal and maximal fixed points of f in $[\bar{y},\hat{y}]$. Let $w = \bar{s} - \hat{s}$.

$$-w'' = \sigma(F(\hat{s}) - F(\bar{s})) \geq 0, \qquad \text{since } \bar{s} \leq \hat{s}$$

$$w(0) = w(1) = 0.$$

Thus from the maximum principle for elliptic equations, $w \geq 0$, and we have $\bar{s} \geq \hat{s}$, together with $\bar{s} \leq \hat{s}$, hence $\bar{s} = \hat{s}$.

<u>Proof of Proposition 1.4</u> - This proof follows a method employed by Pao in [27] to analyze the behavior of immobilized enzyme systems as $t \to \infty$. Let $t_o > 0$. We know that $\phi = s(\cdot,t_o) \in H^1(\Omega)$, $\Omega =]0,1[$, so that $\phi \in C^0([0,1])$. Therefore:

$$s_e - \rho w \leq \phi \quad \text{for} \quad \rho \geq \sup_{0 \leq x \leq 1} \frac{s_e(x) - \phi(x)}{w(x)} .$$

Let us define:

$$(1.70) \qquad \bar{y} = s_e(x) - \rho w(x) e^{-\mu(t-t_o)}$$

and let us check that, on the time interval $[t_o,+\infty[$, $\bar{y}$ is a lower solution for the problem:

$$(1.71) \qquad \begin{cases} s_t - s_{xx} + \sigma F(s) = 0, & 0 < x < 1, \quad t > t_o, \\ s(0,t) = \alpha, \quad s(1,t) = \beta, & \\ s(x,t_o) = \phi(x). & \end{cases}$$

We immediately have:

$$\begin{cases} \overline{y}_t - \overline{y}_{xx} + \sigma F(\overline{y}) = \sigma(F(\overline{y}) - F(s_e)) \leq 0 \\ \overline{y}(0,t) = \alpha, \quad \overline{y}(1,t) = \beta, \\ \overline{y}(x,t_o) \leq \phi(x), \end{cases}$$

the first inequality following from $\overline{y}_t - \overline{y}_{xx} = \rho\mu w e^{-\mu(t-t_o)} - s_e'' + \rho w'' e^{-\mu(t-t_o)}$, $-w'' = \mu w$ and $-s_e'' + \sigma F(s_e) = 0$.

Similarly, it is easy to check that:

$$(1.72) \qquad \hat{y} = s_e(x) + \rho w(x) e^{-\mu(t-t_o)}$$

is an upper solution for (1.71) on the time interval $[t_o,+\infty[$. In fact we even have $s(x,t) \leq s_e(x)$ since s_e is an upper solution for problem (1.50). Therefore Proposition 1.4 follows from an application of Proposition 1.1 to problem (1.71) with $\overline{y}$ and $\hat{y}$ defined by (1.70) and (1.72), and:

$$r = \rho e^{\mu t_o}.$$

Remark 1.7 - The main biochemical consequence of the above numerical and mathematical results is the existence of concentration profiles in enzyme membranes. It is very important to take into account this fact when explaining the behavior of biological membranes: interacting diffusion and reaction may cause substrate (resp. product) concentrations to be much lower (resp. higher) within membranes than at boundaries. And in any case, one cannot speak of a single internal concentration for each chemical species involved, but rather of a concentration profile. A modification of the environment causes the concentration profiles to undergo a fast transient evolution, during a period of the order of L^2/D_S, the diffusion characteristic time, and to approach exponentially a stable steady state.

References

[1] Henri, V., Lois Générales de l'Action des Diastases (Hermann, Paris, 1903).

[2] Michaelis, L. and Menten, M., Die Kinetik der Invertinwirkung, Biochem. Z 49 (1913) 333-369.

[3] Briggs, G.E. and Haldane, J.B.S., A note on the kinetics of enzyme action, Biochem. J. 19 (1925) 338-339.

[4] Heineken, F.G., Tsuchiya, H.M., and Aris, R., On the mathematical status of the pseudo-steady state hypothesis of biochemical kinetics, Math. Biosci. 1 (1967) 95-113.

[5] Murray, J.D., Lectures on Nonlinear Differential-Equation Models in Biology (Clarendon Press, Oxford, 1977).

[6] Fife, P.C., Mathematical Aspects of Reacting and Diffusing Systems (Springer-Verlag, Berlin, 1979).

[7] Banks, H.T., Modeling and Control in the Biomedical Sciences (Lecture Notes in Biomathematics N°6, Springer Verlag, Berlin, 1975).

[8] Walter, C., Contributions of enzyme models, in: Solomon, D.L. and Walter, C. (eds.), Mathematical Models in Biological Discovery (Lecture Notes in Biomathematics N°13, Springer-Verlag, Berlin , 1977).

[9] Bernhard, S.A., The Structure and Function of Enzymes, (Benjamin, New York, 1968).

[10] Chibata, I., Immobilized Enzymes, Research and Development, (Halsted Press, Wiley, New York, 1978).

[11] Broun G., Thomas d., Gellf G., Domurado D., Berjonneau A.M., and Guillon C., New methods for binding enzyme molecules into a water insoluble matrix : properties after insolubilization, Biotechnol. Bioeng., Vol. 15 (1973) 359-375.

[12] Thomas, D., Broun G., and Selegny E., Monoenzymatic model membranes : diffusion-reaction kinetics and phenomena, Biochimie, 54 (1972) 229-244.

[13] Thomas D., Bourdillon C., Broun G., and Kernevez J.P., Kinetic behavior of enzymes in artificial membranes. Inhibition and reversibility effects, Biochemistry, 13 (1974) 2995-3000.

[14] Calvot, C., Berjonneau A.M., Gellf G., and Thomas D., Magnetic enzyme membranes as active elements of electrochemical sensors. Specific amino acid enzyme electrodes, FEBS Letters Vol. 59 (1975) 258-262.

[15] Cordonnier, M., Lawny F., Chapot D., and Thomas D., Magnetic enzyme membranes as active elements of electrochemical sensors. Lactose, saccharose, maltose bienzyme electrodes, FEBS Letters, Vol. 59, N°2 (1975) 263-267.

[16] Durand, P., David A., and Thomas D., An enzyme electrode for acetylocholine, Biochimica et Biophysica Acta, 527 (1978) 277-281.

[17] Richtmyer, R.D. and Morton K.W., Difference Methods for Initial Value Problems (Wiley-Interscience, New York, 1957).

[18] Barbotin, J.N. and Thomas D., Electron microscopic and kinetic studies dealing with an artificial enzyme membrane, Application to a cytochemical model with the horseradish peroxidase-3,31-diaminobenzidine system, J. Hystochem. Cytochem. Vol. 22, N°11 (1974) 1048-1059.

[19] Barbotin, J.N., Electron microscopy as a tool for studying artificial enzyme membranes, in: Thomas, D. and Kernevez J.P. (eds.), Analysis and Control of Immobilized Enzyme Systems (North-Holland, Amsterdam, 1976).

[20] Malpièce, Y., Sharan M., Barbotin J.N., Personne P., and Thomas D., A histochemical model dealing with a glucose-oxidase-peroxidase immobilized bienzyme system: The influence of diffusion limitations on histochemical results (submitted for publication).

[21] Barbotin, J.N., Properties of lactate dehydrogenase immobilized in a lipid-protein matrix, FEBS Letters, Vol. 72, N°1 (1976) 93-97.

[22] Barbotin, J.N. and Thomasset, Immobilization of L-glutamate dehydrogenase into soluble cross-linked polymers - ADP effects and electron microscopy studies, Biochem. Biophys. Acta, 570 (1979) 11-21.

[23] Auchmuty, J.F.G., Qualitative effects of diffusion in chemical systems, in: Lectures on Mathematics in the Life Science, Vol. 10 (The American Mathematical Society, Providence, 1978).

[24] Sattinger, D.H., Monotone methods in nonlinear elliptic and parabolic boundary value problems, Indiana Univ. Math. Jour. 21 (1972) 979-1000.

[25] Sattinger, D.H., Topics in Stability and Bifurcation Theory (Lecture Notes in Mathematics N°309, Springer-Verlag, Berlin, 1973).

[26] Amann, H., Nonlinear operators in ordered Banach spaces and some applications to nonlinear boundary value problems, (preprint).

[27] Pao, C.V., Mathematical analysis of enzyme-substrate reaction diffusion in some biochemical systems, J. Nonlinear Analysis: Theory, Methods, and Applications (to appear).

[28] Lions, J.L., Quelques Méthodes de Résolution des Problèmes aux Limites Non-linéaires (Dunod, Paris, 1969, English translation by Le Van, 1971).

[29] Protter, M.H. and Weinberger H.F., Maximum Principles in Differential Equations (Prentice Hall, Englewood Cliffs, New Jersey, 1967).

CHAPTER 2

GLUCOSE PUMP

Suppose that two compartments I and II, containing solutions of substrate at concentrations S_1 and S_2 ($S_1 < S_2$), are separated by an inactive membrane, namely a membrane prepared as described in Chapter 1, but containing only inactive protein, without enzyme. There is no reaction within the membrane and the only phenomenon taking place in this system is a flow of substrate from compartment II to compartment I due to the diffusion of substrate across the membrane from regions at high concentration to regions at low concentration. Moreover, if concentrations S_1 and S_2 are not held fixed, their difference $S_2 - S_1$ tends towards 0 as $t \to \infty$. Thomas et.al. ([1],[2],[3]) describe a membrane system able to transport glucose against the concentration gradient, namely from compartment I at low concentration S_1 to compartment II at high concentration S_2 (Fig. 2.1). In this chapter we first describe this so-called "active transport" phenomenon (section 2.1), then model this membrane system by a pair of coupled partial differential equations governing S_1 and S_2 concentrations (section 2.2), then give numerical methods and results (section 2.3) and finally discuss mathematical aspects (section 2.4).

2.1 Description

The membrane is made of two enzyme layers sandwiched between two selective films (Fig. 2.2). The enzyme layers contain respectively hexokinase (E_1) and phosphatase (E_2). Hexokinase catalyzes the phosphorylation of glucose (S) into glucose-6-phosphate (P), and phosphatase catalyzes the dephosphorylation of glucose-6-phosphate into glucose:

$$S + ATP \xrightarrow{E_1} P + ADP, \quad P \xrightarrow{E_2} S + \text{inorganic phosphate}.$$

In layer one $S + E_1 \to P + E_1$, while in layer two $P + E_2 \to S + E_2$. The membrane is impregnated with ATP. The selective films are permeable to S, but impermeable to P. The membrane separates two compartments I and II containing solutions of S at concentrations S_1 and S_2 ($S_1 \leq S_2$) (Fig. 2.1). In the experiments reported by Thomas the first (donor) compartment is large enough with respect to membrane volume for S concentration within it (S_1) to be assumed constant, while the second (receptor) compartment is 150 times smaller, and S concentration within it (S_2) is increasing (though $S_2 \geq S_1$ at the beginning of the experiment). Thus this unsymmetrical membrane is a model of a metabolic and spatial sequential enzyme system transporting glucose against the concentration gradient. This phenomenon, called active transport can be explained by examining the S and P concentration profiles, obtained by numerical simulation (Fig. 2.3). S enters from compartment I into layer 1 where it is consumed, producing P. In layer 1, the S profile is convex. P produced in layer 1 diffuses into the second layer where it is converted to S. When the S profile presents a hump in layer 2 as depicted in Fig. 2.3, S leaves the membrane into compartment II where it is found at increasingly higher levels of concentration. This system behaves as a glucose pump where energy is provided by ATP.

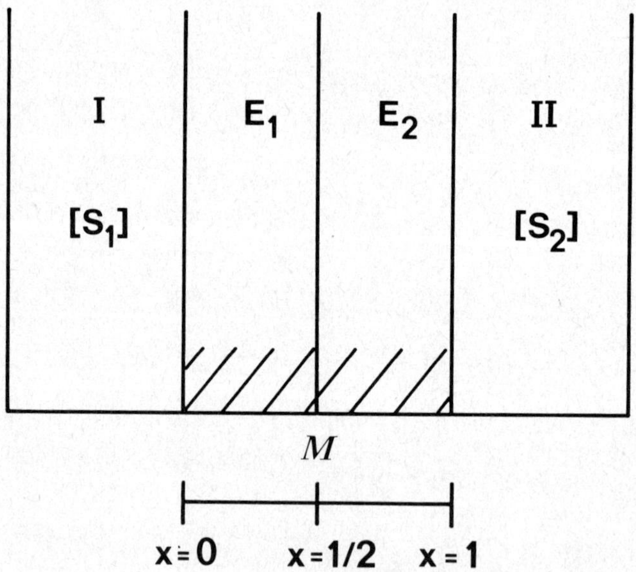

Fig. 2.1 - Double-layer membrane separating 2 solutions of substrate S at concentrations S_1 and S_2 in compartments I and II. In layer 1, containing enzyme E_1, $S + E_1 \rightarrow P + E_1$. In layer 2, containing enzyme E_2, $P + E_2 \rightarrow S + E_2$.

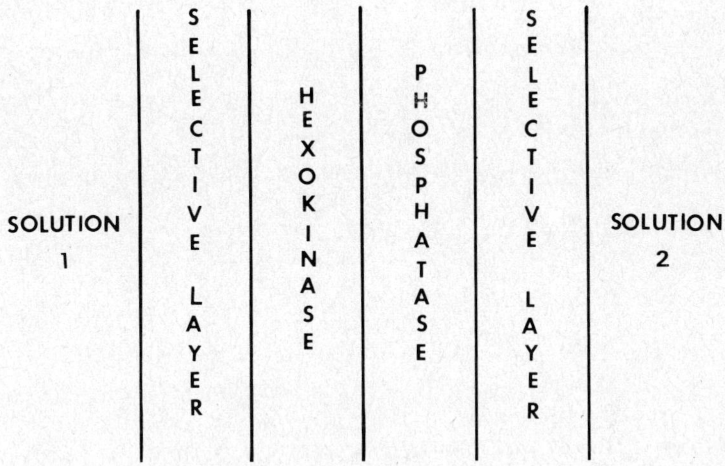

Fig. 2.2 - Diagram of a double-layer membrane sandwiched between two selective layers.

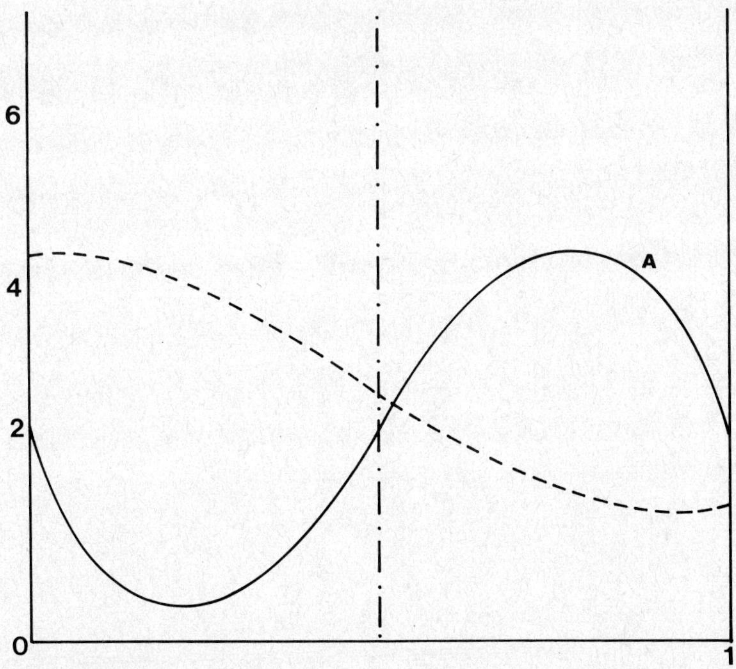

Fig. 2.3 - S and P concentration profiles in the double layer membrane, indicated respectively by solid A for S and the dashed line for P.

2.2 Modeling

The reaction $S + E_1 \rightarrow P + E_1$ is assumed monoenzymatic irreversible with competitive inhibition by the product P. Thus the Briggs-Haldane reaction velocity in layer one is given by:

$$v_1 = \left(\frac{\partial P}{\partial t} \right)_{\text{reaction}} = V_{M_1} S/(K_S(1 + P/K_I) + S).$$

The reaction $P + E_2 \rightarrow S + E_2$ being monoenzymatic irreversible, we have in layer 2 the reaction velocity:

$$v_2 = \left(\frac{\partial S}{\partial t} \right)_{\text{reaction}} = V_{M_2} P/(K_P + P).$$

S and P concentrations within the two active layers are governed by equations:

$$(2.1) \quad \frac{\partial S}{\partial t} - D_S \frac{\partial^2 S}{\partial x^2} + v = 0 \quad \text{and} \quad \frac{\partial P}{\partial t} - D_P \frac{\partial^2 P}{\partial x^2} - v = 0,$$

where the x-axis has been chosen perpendicular to the membrane surfaces, S and P have the same diffusion coefficients D_S and D_P, $v = v_1$ in layer one and $v = -v_2$ in layer two, with $V_{M_1} = V_{M_2} = V_M$. The selective film's thickness in negligible with respect to the overall thickness L of the active layers. Let θ denote the characteristic time for diffusion: $\theta = L^2/D_S$. If we introduce the non-dimensional variables:

$$s = S/K_S, \ p = P/K_S, \ x' = x/L \text{ and } t' = t/\theta,$$

substitute them into (2.1) and for convenience drop the primes, we get the dimensionless <u>glucose pump equations</u>:

$$(2.2) \quad \begin{aligned} & s_t - s_{xx} + F(\cdot,s,p) = 0, \quad 0 < x < 1,\ t > 0, \\ & p_t - p_{xx} - F(\cdot,s,p) = 0, \end{aligned}$$

where the <u>reaction term</u> F is defined by:

$$(2.3)\quad F(x,s,p) = \begin{cases} \sigma s/(1 + |s| + \lambda|p|) & \text{if } 0 < x < 1/2 \\ -\sigma p/(1 + |p|) & \text{if } 1/2 < x < 1, \end{cases}$$

$$(2.4)\quad \sigma = (V_M/K_S)L^2/D_S \quad \text{and} \quad \lambda = K_S/K_I.$$

In (2.2) $F(\cdot,s,p)$ denotes the function $(x,t) \to F(x,s(x,t),p(x,t))$, $0 < x < 1$, $t > 0$.

<u>Initial conditions</u> - The membrane being initially void of S and P, the initial conditions are:

$$(2.5)\quad s(\cdot,0) = p(\cdot,0) = 0,$$

where $s(\cdot,0)$ is the function $x \to s(x,0)$, $0 < x < 1$.

<u>Boundary conditions</u> - Both walls of the membrane being impermeable to P, appropriate boundary conditions for P are:

$$(2.6)\quad p_x(0,\cdot) = p_x(1,\cdot) = 0.$$

Here $p_x(0,\cdot)$ denotes the function $t \to p_x(0,t)$, $t > 0$.

The boundary condition for S at $x = 0$ is:

$$(2.7)\quad s(0,\cdot) = \alpha$$

where $\alpha = S_1/K_S$ either is held fixed by some chemostat, or may be considered as a constant because the donor compartment is large enough.

The boundary condition for S at $x = 1$ is:

$$(2.8)\quad \begin{cases} s_t(1,t) + cs_x(1,t) = 0 \\ s(1,0) = \beta, \end{cases}$$

where $\beta = S_2/K_S$ is the initial concentration in the receptor compartment and:

$$(2.9)\quad c = \frac{v_M}{v_R} = \frac{\text{membrane volume}}{\text{receptor compartment volume}} .$$

Condition (2.8) may be obtained by the following reasoning: the rate of increase of the total amount of S in the receptor compartment, $v_R s_t(1,t)$, equals the total amount of S leaving the membrane per unit of time, $-As_x(1,t)$, A being the membrane area. As v_M = thickness x area = 1 x A = A, equation (2.7) results, with c given by (2.9).

In fact $c \ll 1$ since the membrane volume v_M is small compared to the receptor compartment volume v_R.

We may distinguish two epochs in the evolution of the system. In a first epoch, of small duration, the boundary value $s(1,t)$ has no time to change appreciably, while the S and P profiles of concentration, $s(\cdot,t)$ and $p(\cdot,t)$, undergo a fast transition from initial state (2.5) towards a steady state satisfying:

$$(2.10)\quad \begin{cases} -s''(x) + F(x,s(x),p(x)) = 0, \quad s(0) = \alpha,\ s(1) = \beta, \\ -p''(x) - F(x,s(x),p(x) = 0,\ p'(0) = p'(1) = 0. \end{cases}$$

This fast transition obeys (2.2), (2.5), (2.6), (2.7) and

$$(2.11)\quad s(1,\cdot) = \beta.$$

Fig. 2.4 shows the S and P concentration profiles during this evolution, and the numerical simulations show that these profiles are unchanging after about t = 10 (remember that the unit of time, $\theta = L^2/D_S$, corresponds to 18 seconds if, as is the case, $L = 5\ 10^{-3}$cm and $D_S = 5\ 10^{-3}\text{cm}^2\text{h}^{-1}$).

Second epoch - For larger time behavior the boundary condition (2.11) for S is no more valid since the flow of S from x = 0 to x = 1 gives rise to an increase of S concentration in the receptor compartment, according to equations (2.8). For example, in Thomas' experiments, glucose concentration is observed in the receptor compartment during several hours. Thus, once a quasi-steady state has been attained through a fast transition, this quasi-steady state is going to change slowly during a long period, this evolution being governed by equations (2.2), (2.5), (2.6), (2.7), and (2.8). The numerical simulations give the results shown in Fig. 2.5 and end with steady state profiles as depicted in Fig. 2.6 and which are represented by equations:

$$(2.12)\begin{cases} -s'' + F(\cdot,s,p) = 0, \quad s(0) = \alpha, \quad s'(1) = 0, \\ -p'' - F(\cdot,s,p) = 0, \quad p'(0) = p'(1) = 0. \end{cases}$$

Note that when (2.12) is satisfied there is no more pumping action, since $s'(1) = 0$.

To summarize the above discussion, the modeling of the glucose pump system results in evolution problems governed by equations (2.2), with initial conditions (2.5), supplemented with boundary conditions (2.6) for P, (2.7) for S at $x = 0$, and either (2.8) or (2.11) for S at $x = 1$, and corresponding steady state problems (2.10) or (2.12).

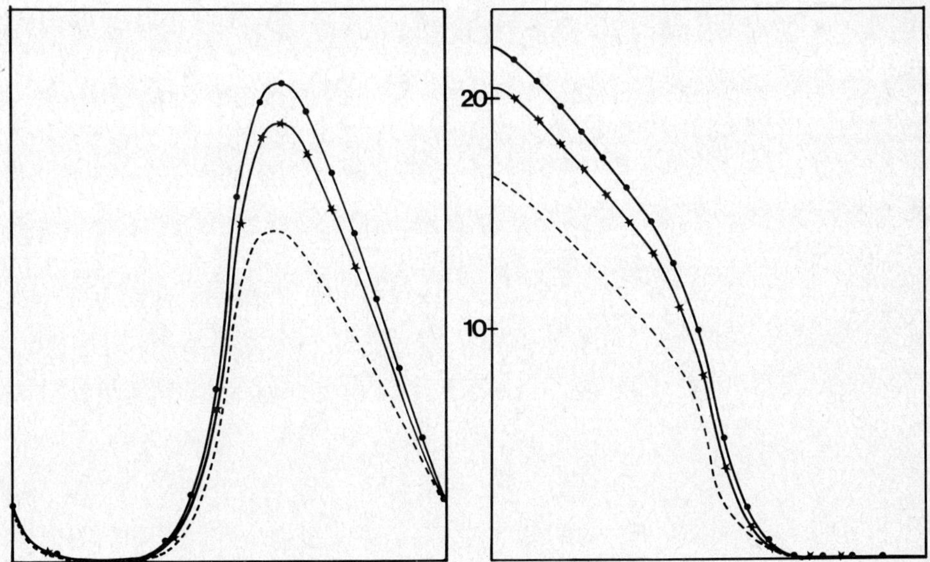

Fig. 2.4 - Evolution of S (on the left) and P (on the right) concentration profiles for equations (2.1), with $\alpha = \beta = 1$ and $\sigma = 700$. Profiles ---, —★—, and —•— correspond respectively to times $t = 0.5$, $t = 1$, & $t = 10$.

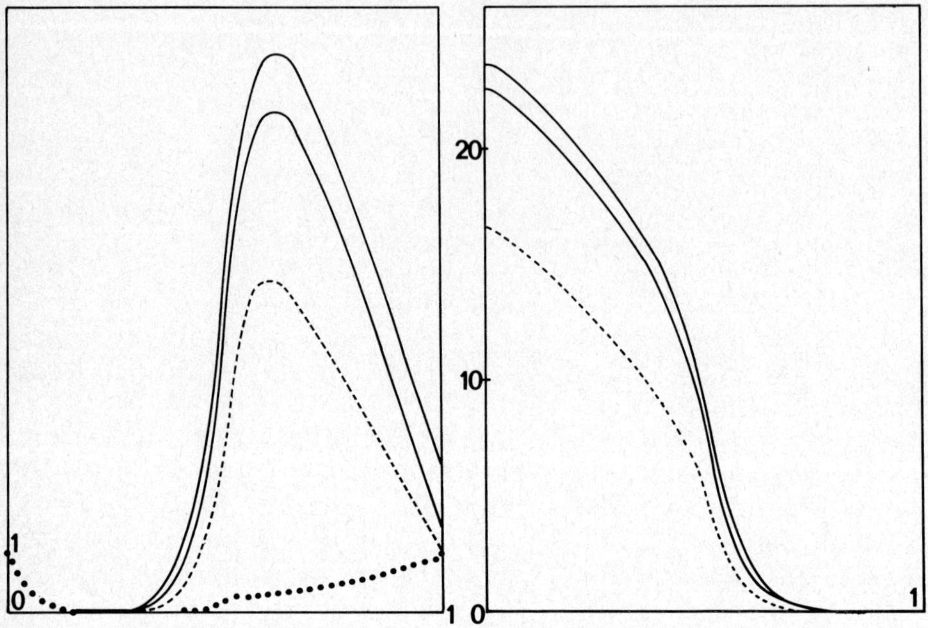

Fig. 2.5 - Evolution of S (on the left) and P (on the right) concentration profiles for equations (2.1) with B.C.(2.2), with $\alpha = \beta = 1$, $\sigma = 700$ and $c = 0.005$. Profiles ····, ----, ——, & —— correspond respectively to times $t = 0.05$, $t = 0.5$, $t = 5$, and $t = 15$.

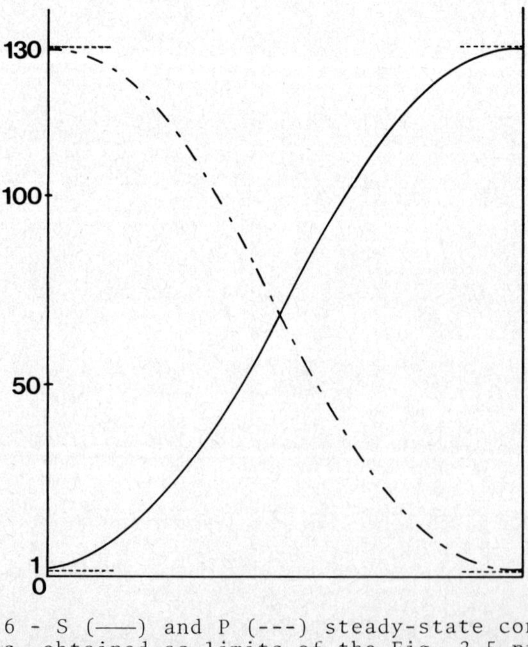

Fig. 2.6 - S (——) and P (---) steady-state concentration profiles, obtained as limits of the Fig. 2.5 profiles.

2.3 Numerical Simulations

The results of the numerical simulation of equations (2.2),(2.5), (2.6), (2.7), and (2.8) (resp. (2.11)) are given in Figures (2.4) and (2.5). In the first case it appears that the profiles ($s(\cdot,t)$ and $p(\cdot,t)$) tend increasingly towards a steady state (s_e,p_e), solution of (2.10), and, after $t = 10$, the approximations of ($s(\cdot,t)$, $p(\cdot,t)$) vary no more with time. $\infty = 10$ for this system !

In the second case, on the contrary, it takes a longer time for the receptor compartment to attain a stationary value. This stationary value is about 130 times the concentration of S in the donor compartment (Fig. 2.6). These numerical simulations aid understanding why unicellular animals immersed in a solution of glucose at low concentration S_1, continue to absorb glucose, even though glucose concentration within them, S_2, is much higher: their membrane could be the seat of similar diffusion-reaction phenomena.

There is little to say about the numerical methods used to obtain these curves in addition to what was said in Section 1.5 on finite difference methods. Again two schemes were employed, the explicit one and the totally implicit one, giving the same results. A stability condition for the explicit scheme being $k = \Delta t < 1/(\sigma + 2/h^2) = 1/(700 + 722)$ if $h = \Delta x = 1/19$, we took $k = 10^{-4}$, and 90,000 time steps were necessary in the first case (Fig. 2.4) to attain a steady state, at about $t = 10$. On the other hand we can use an implicit scheme with $k = 10^{-3}$, which gives the same results with only 9,000 time steps.

<u>Remark 2.1</u> - (about the choice of mesh points) As the rate expression $F(x,s,p)$ is not well defined at $x = 1/2$, it is convenient to avoid $x = 1/2$ as a mesh point by choosing an odd number of intervals in $[0,1]$ ($N = 2M + 1$, $Nh = 1$). Thus the mesh points will be either in layer one ($x = 0, h, \ldots, Mh$) or in layer two ($x = (M+1)h, \ldots, (2M+1)h$) where the rate expression is well defined.

<u>Remark 2.2</u> - (about discretization of boundary condition (2.8)) With a totally implicit finite difference scheme, (2.8) can be discretized by:

$\underline{\bar{y} \leq f(\bar{y})}$ is obvious since, if $f(\bar{y}) = (s,p)$, $s \geq 0$ and $p \equiv 0$.

$\underline{f(\hat{y}) \leq \hat{y}}$ results from the fact that $\hat{y}$ is an upper solution for problem (2.13)-(2.16), in the sense that:

$$(2.24)\quad \begin{cases} q_t - q_{xx} + F(\cdot,q,r) \geq 0, \\ r_t - r_{xx} - F(\cdot,q,r) \geq 0, \\ q(0,\cdot) \geq \alpha,\ q(1,\cdot) \geq \beta, \\ -r_x(0,\cdot) \geq 0, \quad r_x(1,\cdot) \geq 0, \\ q(\cdot,0) \geq 0, \quad r(\cdot,0) \geq 0. \end{cases}$$

Indeed, we have, on $0 < x < 1/2$,

$$\begin{cases} q_t - q_{xx} + F(\cdot,q,r) = \sigma + \sigma \dfrac{q}{1 + q + \lambda r} > 0, \\ \\ r_t - r_{xx} - F(\cdot,q,r) = a - \sigma \dfrac{q}{1 + q + \lambda r} \geq 0, \end{cases}$$

and, on $1/2 < x < 1$,

$$\begin{cases} q_t - q_{xx} + F(\cdot,q,r) = \sigma - \sigma \dfrac{r}{1 + r} > 0, \\ \\ r_t - r_{xx} - F(\cdot,q,r) = -a + \sigma \dfrac{r}{1 + r} \geq -a + \dfrac{\sigma r(1)}{1 + r(1)} = \\ \\ \qquad\qquad -a + \sigma \dfrac{m}{1 + m} \geq 0. \end{cases}$$

The boundary and initial conditions for (q,r) to be an upper solution are trivially satisfied. Let $f(\hat{y}) = (s,p)$, defined by (2.21) with $u = q$ and $v = r$. Subtracting these equations from (2.24) we obtain, for $w = q - s$ and $z = r - p$,

$$\begin{cases} w_t - w_{xx} + \sigma w \geq 0, \quad w(0,\cdot) \geq 0,\ w(1,\cdot) \geq 0,\ w(\cdot,0) \geq 0, \\ z_t - z_{xx} + \sigma z \geq 0,\ -z_x(0,\cdot) \geq 0,\ z_x(1,\cdot) \geq 0,\ z(\cdot,0) \geq 0 \end{cases}$$

whence $w \geq 0$ and $z \geq 0$ from the Maximum Principle for parabolic problems. But $(w,z) \geq 0$ means $\hat{y} \geq f(\hat{y})$.

f is increasing - Suppose $(u_1,v_1) \leq (u_2,v_2)$ and let $w = s_2 - s_1$, $z = p_2 - p_1$, where $(s_i,p_i) = f(u_i,v_i)$, $i = 1,2$. We have:

$$(2.25)\begin{cases} w_t - w_{xx} + \sigma w = \sigma(u_2-u_1) - F(\cdot,u_2,v_2) + F(\cdot,u_1,v_1), \\ w(0,\cdot) = 0, \quad w(1,\cdot) = 0, \quad w(0,\cdot) = 0, \end{cases}$$

$$(2.26)\begin{cases} z_t - z_{xx} + \sigma z = \sigma(v_2-v_1) + F(\cdot,u_2,v_2) - F(\cdot,u_1,v_1), \\ z_x(0,\cdot) = z_x(1,\cdot) = 0, \quad z(\cdot,0) = 0. \end{cases}$$

The second member in (2.25) can be written:

$$[\sigma(u_2-u_1) - F(\cdot,u_2,v_2) + F(\cdot,u_1,v_2)] + [F(\cdot,u_1,v_1) - F(\cdot,u_1,v_2)]$$

where the quantities within brackets are positive, the first one because of the Lipschitz continuity of F, the second one because F is a decreasing function of p. Thus, from the Maximum Principle for parabolic problems, $w \geq 0$. Similarly, the second member in (2.26) is positive, hence $z \geq 0$.

Continuity and compactness of f - f, defined by equations (2.21), may be considered as built up from three mappings. The first one transforms (u,v) in (Q,R) and is continuous from E to E, as follows from Lipschitz continuity of F. The second one transforms (Q,R) in (u,v), and is continuous from E to $[\phi + W(0,T)] \times W_1(0,T)$, where $\phi = \alpha(1 - x) + \beta x$,

$$W(0,T) = \{\phi \mid \phi \in L^2(0,T;H^1_0(\Omega)),\ \phi_t \in L^2(0,T;H^{-1}(\Omega)\},\ \Omega =]0,1[,$$

and

$$W_1(0,T) = \{\psi \mid \psi \in L^2(0,T;H^1(\Omega)),\ \psi_t \in L^2(0,T;(H^1(\Omega))')\}.$$

Finally, the third one is the natural injection $(s-\phi,p) \to (s-\phi,p)$ from $W(0,T) \times W_1(0,T)$ to E, and is compact.

iii/ $t \to (s(\cdot,t),p(\cdot,t))$ is an increasing function of time - Let $u(x,t) = s(x,t+\tau)$ and $v(x,t) = p(x,t+\tau)$, where (s,p) is the solution of problem (2.13)-(2.16) and $\tau > 0$. u and v satisfy the same equations and boundary conditions as s and p, and initial conditions $u(x,0) = s(x,\tau) > 0$, $v(x,0) = p(x,\tau) > 0$. Hence (u,v) is an upper solution for problem (2.13)-(2.16). Then, replacing $\hat{y}$ by (u,v) in the above argument, it follows that $(s,p) \leq (u,v)$, which means that $s(\cdot,t) \leq s(\cdot,t+\tau)$ and $p(\cdot,t) \leq p(\cdot,t+\tau)$.

iv/ <u>$(s(\cdot,t),p(\cdot,t))$ tends towards a steady state when $t \to \infty$</u> : as $(s(\cdot,t),p(\cdot,t)) \leq \hat{y} = (q,r)$, for every x the functions $t \to s(x,t)$ and $t \to p(x,t)$ tend towards limits $u(x)$ and $v(x)$. Moreover, from the Lebesgue theorem, $s(\cdot,t) \to u$ in $L^2(\Omega)$ and $p(\cdot,t) \to v$ in $L^2(\Omega)$, whence, by using the Lipschitz continuity of F, $F(\cdot,s(\cdot,t),p(\cdot,t)) \to F(\cdot,u,v)$ in $L^2(\Omega)$. Let w and z be the solutions of:

$$\begin{cases} -w''(x) = -F(x,u(x),v(x)), & 0 < x < 1, \\ w(0) = \alpha, \quad w(1) = \beta \end{cases}$$

and

$$\begin{cases} -z''(x) + \sigma z(x) = \sigma v(x) + F(x,u(x),v(x)), & 0 < x < 1, \\ z'(0) = z'(1) = 0. \end{cases}$$

We wish to prove that $(w,z) = (u,v)$. Let us define $y(x,t) = w(x) - s(x,t)$. We have:

$$(2.27)\begin{cases} y_t - y_{xx} = f(x,t), & 0 < x < 1, \quad t > 0, \\ y(0,t) = 0, \quad y(1,t) = 0, \\ y(x,0) = w(x) \end{cases}$$

where

(2.28) $f(\cdot,t) \to 0$ in $L^2(\Omega)$ as $t \to \infty$,

since $f(\cdot,t) = F(\cdot,s(\cdot,t),p(\cdot,t)) - F(\cdot,u,v)$.

We need the following Lemma:

<u>Lemma 2.1</u> - Let y satisfy (2.27) and (2.28). Then $y(\cdot,t) \to 0$ in $L^2(\Omega)$ as $t \to \infty$.

Proof - Let $\varepsilon > 0$ be given, arbitrarily small. For t large enough, we have $|f(\cdot,t)| < \varepsilon$, where $|\cdot|$ denotes the $L^2(\Omega)$ - norm. If in (2.27) we multiply by y, and integrate with respect to x on (0,1), we obtain:

$$\frac{1}{2}\frac{d}{dt}|y(\cdot,t)|^2 + |y_x(\cdot,t)|^2 = \int_0^1 f(x,t)y(x,t)dx$$

$$\leq |f(\cdot,t)|\ |y(\cdot,t)| \leq \varepsilon\ |y(\cdot,t)| \leq \frac{1}{2}\varepsilon^2 + \frac{1}{2}|y(\cdot,t)|^2.$$

Applying Poincaré Inequality $|y_x(\cdot,t)|^2 \geq |y(\cdot,t)|^2$ and denoting $\phi(t) = |y(\cdot,t)|^2$, we have $\phi'(t) + \phi(t) \leq \varepsilon^2$, which implies $\phi(t) \leq \varepsilon^2 + (\phi(0) - \varepsilon^2)e^{-t}$. Thus for t large enough $\phi(t) \leq 2\varepsilon^2$.

Applying lemma 2.1 to our problem, it results that $s(\cdot,t) \to w$ as $t \to \infty$, hence $u = w$. Similarly, $p(\cdot,t) \to z$ as $t \to \infty$, and $v \equiv z$. Thus u and v satisfy the steady state equations (2.18)-(2.19). The solutions of these equations are in $H^1(\Omega)$, hence in $C^0(\overline{\Omega})$. On the other hand, for $t > 0$, $s(\cdot,t)$ and $p(\cdot,t)$ belong to $H^1(\Omega)$, hence to $C^0(\overline{\Omega})$. By the Dini theorem [6] $s(\cdot,t) \to u$ in $C^0(\overline{\Omega})$, and $p(\cdot,t) \to v$ in $C^0(\overline{\Omega})$.

Remark 2.3 - The above argument demonstrates the existence of at least one steady-state, solution of (2.18)-(2.19). It is possible to prove existence of minimal and maximal solutions of (2.18)-(2.19) in $[\overline{y},\hat{y}]$, where $\overline{y} = (0,0)$ and $\hat{y} = [p,q]$ as defined by (2.22) and (2.23), by applying Theorem 1.1 to the situation where $E = (L^2(\Omega))^2$, $P = \{(u,v)\,|\,u(x) \geq 0 \text{ and } v(x) \geq 0 \text{ a.e.}\}$ and $f : (u,v) \to (s,p)$ is defined by:

$$(2.29)\quad \begin{cases} -s'' + \sigma s = \sigma u - F(\cdot,u,v), \\ s(0) = \alpha,\ s(1) = \beta, \end{cases}$$

and

$$(2.30)\quad \begin{cases} -p'' + \sigma p = \sigma v + F(\cdot,u,v), \\ p'(0) = p'(1) = 0. \end{cases}$$

The details are similar to those found in the proofs of propositions

1.2 and 2.1. Let $(\bar{s},\bar{p})$ and $(\hat{s},\hat{p})$ be these minimal and maximal solutions. Whether they are the same or not is an open problem. Numerically, it is possible to approximate $(\bar{s},\bar{p}) = \bar{x}$ and $(\hat{s},\hat{p}) = \hat{x}$ by algorithms using the fact that $f^k(\bar{y}) \to \bar{x}$ and $f^k(\hat{y}) \to \hat{x}$ (Theorem 1.1), namely:

$$\begin{cases} x^0 = \bar{y} \text{ or } \hat{y} \\ x^{k+1} = f(x^k), \end{cases}$$

or, more precisely,

. start either from $s^0 = 0$, $p^0 = 0$, or from $s^0 = q$, $p^0 = r$,
. define s^{k+1} and p^{k+1} as solutions of (2.29) and (2.30) with $u = s^k$ and $v = p^k$.
. stop when s^k and p^k do not vary appreciably anymore.

At least for the values $\sigma = 700$, $\lambda = 1$, and $\alpha = \beta = 1$, there is strong numerical evidence that $\bar{x} \neq \hat{x}$ (Figure 2.7). Anyway, even if there are multiple steady states, $(s(\cdot,t),p(\cdot,t))$ converges towards the minimal solution of (2.18)-(2.19), $\bar{x} = (\bar{s},\bar{p})$, since $(\bar{s},\bar{p})$ being an upper solution for the evolution problem (2.13)-(2.16), $(s(\cdot,t),p(\cdot,t)) \leq (\bar{s},\bar{p}) \leq (u,v)$. There was only one steady state for the simple biochemical system considered in Chapter 1. Here, though it seems probable that there exists only one steady state, we did not find any method to prove it. In subsequent chapters we shall deal with systems possessing multiple steady states. In fact, it is the general situation: initial value problems, governed by equations such as $\frac{dy}{dt} + G(y) = 0$, $y(0) = y_o$, admit naturally only one evolution, whereas the corresponding steady states, obeying $G(y) = 0$, may be multiple.

We proved Proposition 2.1 in the case of boundary conditions (2.16) and (2.19) for s. In the case of boundary conditions (2.17) and (2.20), the reasoning is the same as above, with only two slight changes, relative to the definition of s and to the definition of an upper solution $\hat{y}$. These modifications are explained in Remarks 2.2 and 2.3.

Remark 2.4 - Consider the problem:

$$(2.31)\quad \begin{cases} s_t - s_{xx} + \sigma s = f(x,t), \quad 0 < x < 1, \quad 0 < t < T, \\ s(0,t) = \alpha, \quad s_t(1,t) + cs_x(1,t) = 0, \\ s(x,0) = 0, \quad s(1,0) = \beta. \end{cases}$$

Without entering into details let us say that the Galerkin method can be used to prove existence and uniqueness of a solution for (2.31), the proof being based on the use of the variational formulation:

$$(2.32)\quad \begin{cases} \frac{d}{dt}\left[\int_o^1 s(x,\cdot)v(x)dx + \frac{1}{c}\, s(1,\cdot)v(1)\right] + \int_o^1 (s_x(x,\cdot)v'(x)dx + \\ \int_o^1 s(x,\cdot)v(x)dx = \int_o^1 f(x,t)v(x)dx, \\ \int_o^1 s(x,0)v(x)dx + \frac{1}{c}\, s(1,0)v(1) = \frac{1}{c}\,\beta v(1), \\ \text{for every } v \in H^1(\Omega),\ v(0) = 0. \end{cases}$$

Remark 2.5 - An upper solution $\hat{y}$ for the case of "moving" boundary values for s is $\hat{y} = (q,r)$ such that:

$$q = \alpha + (\sigma/2)x(2-x)$$

and r defined as in (2.23), with:

$$m \geq \tilde{q} = \max_{0 \leq x \leq 1} q(x) \text{ and } \sigma\tilde{q}/(1 + \tilde{q}) \leq a \leq \sigma m/(1 + m)$$

Remark 2.6 - The biochemical interpretation of the above numerical and mathematical results is that two "scalar" phenomena, namely diffusion and reaction, can induce, when interacting, a "vectorial" phenomenon, more precisely an active transport of substrate.

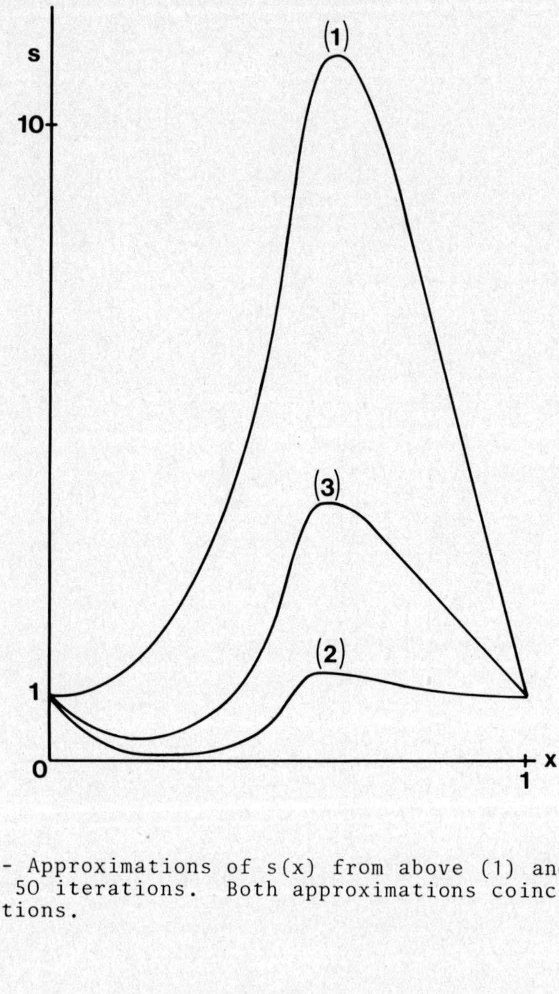

Fig. 2.7 - Approximations of s(x) from above (1) and from below (2) after 50 iterations. Both approximations coincide (3) after 750 iterations.

References

[1] Broun, G., Thomas D., and Selegny E., Structured bienzymatical models formed by sequential enzymes bound into artificial supports: Active glucose transport effect, J. Membrane Biol. 8 (1972) 313-332.

[2] Thomas, D. and Broun, G., Some aspects of the regulation and transport in enzyme membrane models, Biochimie, Tome 55, N°8 (1973) 975-984.

[3] Thomas, D., Artificial enzyme membranes, transport, memory, and oscillatory phenomena, in: Thomas, D. and Kernevez ,J.P. (eds) Analysis and Control of Immobilized Enzyme Systems (North Holland, Amsterdam, 1976).

[4] Amann, H., Nonlinear operators in ordered Banach spaces and some applications to nonlinear boundary value problems, (preprint).

[5] Kernevez, J.P. and Thomas, D., Numerical analysis and control of some biochemical systems, Appl. Math. Optimization, vol. 1, N°3 (1975) 222-285.

[6] Dieudonné , J., Foundations of Modern Analysis (Academic Press, New Jersey, 1960).

CHAPTER 3

MULTIPLE STEADY STATES AND HYSTERESIS

In this chapter we shall be interested in the multiple steady states which may exist in systems involving diffusion and substrate inhibition kinetics.

In Section 3.1 we shall consider the so-called zero-dimensional system, governed by equation:

(3.1) $\frac{ds}{dt} = s_o - s - \rho F(s)$

where:

(3.2) $F(s) = s/(1 + |s| + ks^2)$.

Equation (3.1) models an inactive membrane separating an enzymatically active compartment at concentration s from a reservoir at concentration s_o (Fig. 3.1). Here diffusion and reaction take place respectively within the membrane and the compartment. Though simple, this model can exhibit multiple steady states, governed by the algebraic equation:

(3.3) $s_o - s - \rho F(s) = 0$

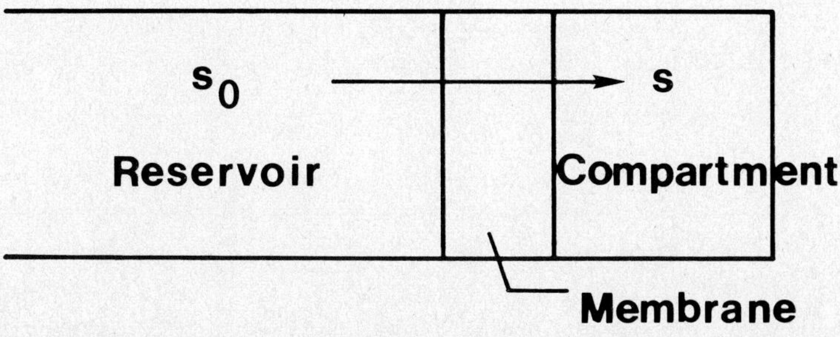

Fig. 3.1

As a consequence we may have an hysteresis phenomenon when, ρ being held fixed, s_o is allowed to vary, first increasingly, then decreasingly. A steady-state s that has been changed by a variation of s_o may fail to return to its original value when the cause of the change is removed.

In Section 3.2 diffusion and reaction take place in the same regions of an artificial enzyme membrane, a one-dimensional system represented by equations:

$$(3.4)\quad \begin{cases} s_t - s_{xx} + \sigma F(s) = 0, & 0 < x < 1, \quad t > 0, \\ s(0,t) = s(1,t) = s_o \end{cases}$$

for the transient evolution and

$$(3.5)\quad \begin{cases} -s''(x) + \sigma F(s(x)) = 0, & 0 < x < 1, \\ s(0) = s(1) = s_o \end{cases}$$

for the stationary states.

Depending upon the values of σ and s_o, the stationary problem (3.5) may admit multiple solutions. This is shown by numerical experiments, and can be demonstrated by elementary methods. We discuss the hysteretic behavior of these solutions when, σ being fixed sufficiently large, s_o varies.

However it appears clearly that in more complex cases a mathematical analysis of similar situations is not within our grasp and that, even in the case of equation (3.5), numerical methods should be used for exploring the family of solutions when some parameter, say s_o, varies. By the family of solutions of (3.5), we mean the subset $\mathcal{A}$ of $R^+ \times B$ defined by:

$$(\lambda, u) \in \mathcal{A} \Leftrightarrow -u''(x) + \sigma F(u(x) + \lambda) = 0, \qquad 0 < x < 1$$

where $B = C_0^2([0,1]) = \{u | u \in C^2([0,1], u(0)) = u(1) = 0\}$.

Accordingly, Section 3.3 describes the continuation method of Kubicek

[1] for problems where the steady-state u of the system depends on a parameter λ through a relation of the form:

$$(3.6) \quad f(\lambda,u) = 0.$$

Here f is a continuously differentiable mapping of $R^+ \times R^n$ into R^n.

Section 3.4 is devoted to a mathematical analysis of the curve of solutions for (3.5) when σ is held fixed and s_o varies. In addition to the familiar methods of upper and lower solutions, which enable us to show the existence of maximal and minimal solutions, we employ the Implicit Function Theorem, thus obtaining continuous curves of solutions. In particular the turning points of the S-shaped curve of solutions (Figure 3.8) are studied by applying results of Crandall and Rabinowitz [2].

In Section 3.5 we discuss the stability of the steady-state solutions of (3.4), showing that the maximal and minimal solutions of (3.5) are stable and that the stability of the steady state solution is changing at turning points. Here again results of [2] are very helpful.

Lastly, Section 3.6 argues the quasi-steady-state system (3.3) as a limit of the evolution system (3.1).

3.1 The Zero-dimensional System

Our model is a well-stirred, enzymatically active compartment separated from a well-stirred reservoir by a membrane (Figure 3.1). The membrane, in which all of the diffusive resistance to mass transfer is concentrated, is inactive, that is without enzyme. It is permeable to substrate, but not to enzyme. The enzyme is uricase, which catalyzes the oxidation of uric acid according to:

$$(3.7) \quad \text{uric acid} + \text{oxygen} \xrightarrow{\text{uricase}} \text{allantoin} + \text{other products}$$

We can assume that oxygen undergoes zero order kinetics so that the only substrate to be considered is in our modelling uric acid. The compartment contains enzyme molecules which catalyze the transformation of this substrate with the rate expression:

$$(3.8) \quad R(S) = V_M\, S/(K_S + S(1 + S/K_{SS})).$$

In this commonly accepted rate law for substrate inhibition kinetics, S is the substrate (uric acid) concentration, while V_M, K_S, and K_{SS} denote the familiar maximal rate, Michaelis, and inhibition constants. The reservoir contains substrate at a fixed concentration S_o. If S is the substrate concentration in the compartment, the membrane transports substrate from the reservoir to the compartment with rate $(D_S\ A/L)(S_o - S)$ where D_S is the substrate diffusion coefficient, A the contact surface area and L the membrane thickness. Mass balance for S in the reactor is expressed by equation:

$$V_r \frac{dS}{dt} = (D_S\ A/L)(S_o - S) - V_r\ R(S)$$

where V_r is the reactor volume. Let P denote the mass transfer coefficient:

$$P = D_S\ A/(L\ V_r) = (V_m/V_r)\ D_S/L^2$$

where V_m is the membrane volume. We have:

$$\frac{dS}{dt} = P(S_o - S) - R(S).$$

Let θ be the characteristic time $\theta = 1/P$. Taking θ and K_S as the natural units of time and concentration we obtain:

$$\frac{ds}{d\tau} = s_o - s - \rho F(s) \tag{3.9}$$

where:

$$s = S_o/K_S,\ \tau = t/\theta,\ \ s_o = S_o/K_S$$

ρ is the reaction-permeation modulus:

$$\rho = (V_M/K_S)\theta, \quad \theta = (V_r/V_m)L^2/D_S,$$

and

$$F(s) = s/(1 + |s| + ks^2) \quad \text{with } k = K_S/K_{SS}. \tag{3.10}$$

Henceforth, in this chapter F(s) will denote the rate expression (3.10). We have chosen to use the simple steady state form for both permeation and reaction in a system which is at least potentially non-stationary. A limit is thereby placed on the maximum frequency for which equation (3.9) will be a good model if s_o is allowed to vary with time. This frequency is typically the order of magnitude P, since diffusion processes relax more slowly than most reactions. In the stationary state, the time derivative in equation (3.9) vanishes, and the mass balance equation reduces to the algebraic equation:

$$(3.11) \quad s_o - s - \rho F(s) = 0.$$

Equation (3.11) is one of the simplest imaginable models in which the coupling of reaction and diffusion yields a phenomenon involving multiple steady states. In fact, it is also an open system, due to transport of substrate from outside into the compartment with a flux $s_o - s$. Without this transport the substrate concentration s would evolve according to the kinetic equation:

$$(3.12) \quad \frac{ds}{d\tau} = - \rho \, F(s)$$

towards the unique equilibrium state s = 0. But equation (3.11) possesses multiple solutions, simply obtained by cutting the curve $y = \rho F(s)$ by the straight line $y = s_o - s$ (Figure 3.2.a). If we fix ρ in (3.11) and wish to obtain s as s_o varies, we can consider the functions:

$$s \rightarrow s_o = s + \rho F(s) \quad \text{and} \quad s \rightarrow F'(s)$$

see (Figures 3.3a and 3.3b) respectively.

It is easily seen that if:

$$\rho > -1/\min_{s \geq 0} F'(s),$$

there are 2 values s^* and s^{**} of s such that:

$$\frac{ds_o}{ds} = 1 + \rho F'(s) = 0,$$

and for which $s_o = s + \rho F(s)$ admits relative extrema. In that case we have the representations depicted in (Figures 3.3c and 3.3d) for s and y as functions of s_o, y being the reaction rate of the system:

(3.13) $y = s_o - s = \rho F(s)$.

The plots given in Figure 3.3 show the existence, for sufficiently large values of ρ, of a multiple steady state region in which three values of the state s are possible for a given external concentration s_o. The multiple stationary state region is bounded by a pair of concentrations s_o^* and s_o^{**} for which the observed stationary state is a discontinuous function of s_o if this parameter is allowed to vary quasi-statically, that is slowly enough for the system to undergo sequential steady states. There is a critically large value of ρ for the onset of this behavior, namely:

$$\rho > -1/ \min_{s \geq 0} F'(s)$$

It is easy to understand the hysteresis phenomenon by observing the sequential stationary states when s_o is allowed to vary slowly, first increasingly (points 1 to 4), then decreasingly (points 4 to 7) as indicated in Figures 3.2b and 3.4:

(Figure 3.2) - Figure 3.2a shows that s is a solution of $s_o - s = \rho F(s)$ if it is the abscissa of a point belonging equally to the curve $y = \rho F(s)$ and the straight line $y = s_o - s$. Figure 3.2b shows those points (s,y) obtained as s_o varies, first increasingly (points 1-4), then decreasingly (points 4-7). Figure 3.4 shows the same states 1-7 on the path followed by the point (s_o,ρ,s) on the catastrophe surface $s_o - s - \rho F(s) = 0$ when ρ is kept fixed and s_o is allowed to vary as in Figure 3.2b.

For $s_o \in]s_o^{**},s_o^*[$ three types of steady states are possible, of which two are stable. In one type of stable state (lying on the upper branch of the S-shaped curve in Figure 3.3c) the reaction is relatively slow and s is close to s_o; these will be designated <u>reaction-controled</u> states. In the other type of stable state (on the lower branch of the S) the reaction is fast and s is limited by diffusion, and may be relatively far from s_o; these will be designated <u>diffusion-controled</u> states.

These notions of reaction and diffusion controled states were introduced by Hardt, Naparstek, Segel, and Caplan for the "papaîn system"[3].

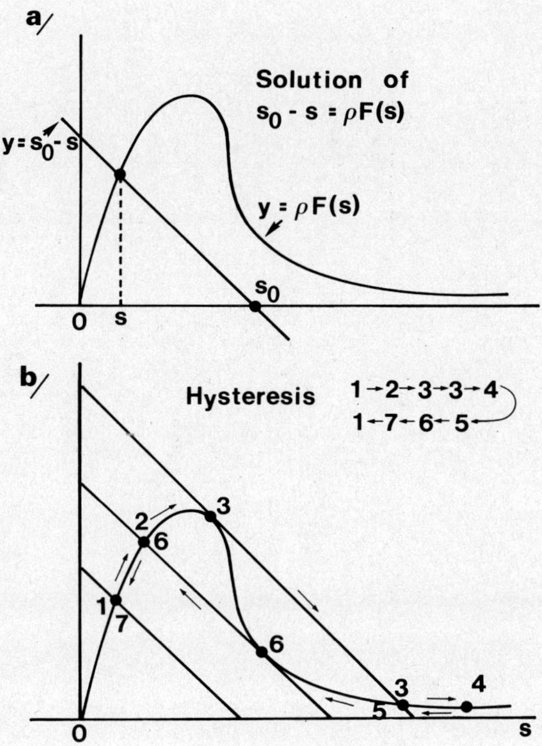
a/
Solution of
s0 - s = ρF(s)
y = s0 - s
y = ρF(s)
s0
0
s
b/
Hysteresis
1→2→3→3→4
1←7←6←5
1
2
3
6
7
6
3
4
5
0
s

Fig. 3.2

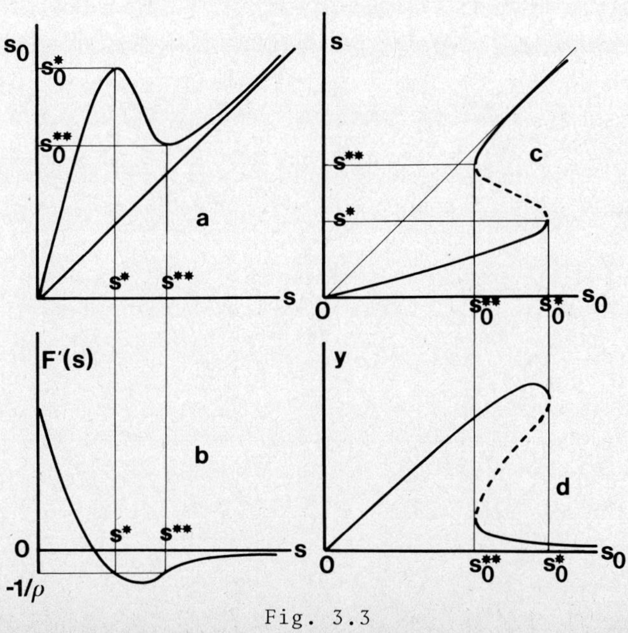
s0
s0*
s0**
a
s*
s**
s
s
s**
c
s*
0
s0**
s0*
s0
F'(s)
b
0
s*
s**
s
-1/ρ
y
d
0
s0**
s0*
s0

Fig. 3.3

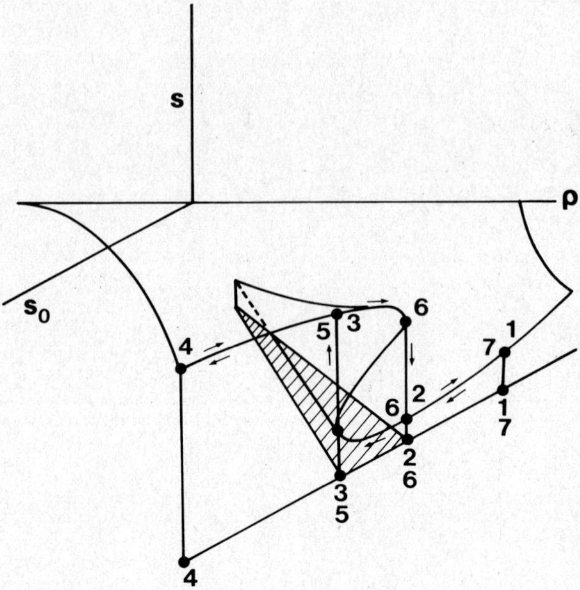

Fig. 3.4

All of the above is easily seen, as is the stability of the state s with respect to the dynamical system (3.1), which changes when $1 + \rho F'(s)$ changes sign, that is at the turning points $s = s^*$ and $s = s^{**}$. However, the behavior of this simple system extends to more complex systems. In particular an S-shaped curve as shown in Figure 3.3c will still represent the qualitative behavior or the distributed system studied in the next section.

3.2 The One-dimensional System

The distributed system to be discussed hereafter is more realistic than the previous one for representing the interaction of diffusion and reaction in living cells. In fact enzyme molecules are mostly located within cell membranes and organelles, that is in those regions where diffusive resistance to mass transfer is concentrated. The so-called one-dimensional system is an artificial enzyme membrane which was experimentally shown to exhibit an hysteretic phenomenon [4].

3.2.1 - Experimental set up: [4] The enzyme is uricase, which catalyzes the oxidation of uric acid according to the stoechiometric equation (3.7). The artificial membrane was prepared by the following procedure: a cross-linking reaction occurred for 2 hours in a solution containing albumin, glutaraldehyde, and uricase. Then this solution was spread on a plane glass surface. After some hours a complete insolubilization occurred and a 50μ-thick proteic membrane was produced. Once the membrane was produced, it was immersed in a solution of uric acid.

The reaction (3.7) needs oxygen as a cosubstrate. To reduce the mass transfer problems of oxygen, several pieces of the same membrane were immersed in a well stirred solution of substrate and put in continuous air bubbling. Though the oxygen concentration in the system is much lower than the uric acid concentration, zero order kinetics for oxygen exists, because the oxidation of uric acid to allantoin is the rate limiting step in the reaction chain. That is to say that the rate expression is given by (3.8) and depends only on S, uric acid concentration.

A time gradient of uric acid was established in the bulk solution by a controled pumping (chemical stat). Thus the concentration of S at the membrane boundaries was changing slowly with time, either increasingly or decreasingly. This variation was slow enough for the membrane to undergo sequential steady states. While uric acid concentration in the bulk solution was thus allowed to vary quasi-statically, the rate of substrate consumption was recorded spectrophotometrically.

It was experimentally demonstrated that enzyme activity, that is the rate of substrate consumption, did present hysteresis when the substrate concentration in the bulk solution was changing.

3.2.2 - Mathematical modeling: Henceforth substrate will mean uric acid, the concentration of which within the membrane obeys:

$$\frac{\partial S}{\partial t} - D_S \frac{\partial^2 S}{\partial x^2} + R(S) = 0,$$

together with boundary conditions:

$$S = S_o \text{ at } x = 0 \text{ and } x = L, \quad (L \text{ membrane thickness}),$$

and initial conditions:

$$S = 0 \text{ at } t = 0 \text{ or } S = S_o \text{ at } t = 0,$$

according to whether the membrane is initially empty or full of substrate. We take as new units K_S for concentration, $\theta (= L^2/D_S)$ for time, L for length, and define the dimensionless quantities $s = S/K_S$, $t' = t/\theta$, and $x' = x/L$. Dropping the primes for convenience, we obtain the following equations:

$$(3.14) \quad \begin{cases} s_t - s_{xx} + \sigma F(s) = 0, & 0 < x < 1, \quad t > 0, \\ s(0,t) = s(1,t) = s_o \\ s(x,0) \quad \text{given} \end{cases}$$

where:

$$(3.15) \quad \sigma = (V_M/K_S)\theta, \quad \theta = L^2/D_S$$

and F(s) is given by (3.10).

We are interested by the steady state, satisfying the 2-point boundary value problem:

$$(3.16) \quad \begin{cases} -s''(x) + \sigma F(s(x)) = 0, & 0 < x < 1, \\ s(0) = s(1) = s_o. \end{cases}$$

If the parameter s_o is allowed to vary with time, the evolution equations (3.14) are more appropriate than the steady state equations (3.16) for representing the system evolution. However, since the rate of change of s_o is small with respect to the speed of the system when it evolves towards equilibrium for a given value of s_o, it is possible to make the so-called quasi-steady-state hypothesis, which consists in considering the system as well represented by the steady state equations (3.16), where time t is a parameter appearing through s_o. The justification of such a quasi-steady-state hypothesis is given in Section 3.6 for the 0-dimensional case.

3.2.3 - Numerical simulations: Let us consider problems (3.14) and (3.16) for a given boundary value s_o.

It is fairly easy to check that both the evolution and stationary problems (3.14) and (3.16) admit 0 and s_o as lower and upper solutions. Therefore, upon applying Propositions 1.1 and 1.2 we already know that the evolution problem (3.14) admits an unique positive solution satisfying $0 \leq s(x,t) \leq s_o$, and the stationary problem (3.16) admits maximal and minimal solutions $\bar{s}$ and $\hat{s}$, such that $0 \leq \bar{s}(x) \leq \hat{s}(x) \leq s_o$. Also, applying the same kind of arguments as in the proof of Proposition 2.1, we can show that if the evolution system starts from the initial condition $s(\cdot,0) = 0$ (resp. $s(\cdot,0) = s_o$), then the profile $s(\cdot,t)$ tends increasingly (resp. decreasingly) towards the steady state $\bar{s}$ (resp. $\hat{s}$).

A numerical simulation of (3.16) when s_o is a slowly varying function of time, first increasing, then decreasing, was performed according to the following procedure. We solved the initial boundary value problem (3.14) for a succession of values s_o, starting at each step from the slightly perturbed steady state of the preceding step. Sequential stable steady state profiles were observed as indicated in Figure 3.5. (The profiles 1-2-3-4-5-6-7 represent $s(x)/s_o$ as a function of x. Low profiles 1-2-3 correspond to increasing s_o values, until a jump to the high profile 3 for $s_o = s_o^{*} \simeq 68$. High profiles 4-5-6 correspond to decreasing s_o values, until a jump to the low profile 6 for $s_o = s_o^{**} \simeq 61$. $\sigma = 1200$, $k = 0.1$).

It can be checked that these steady states are the minimal or maximal solutions of (3.16), which can be obtained by the following numerical scheme:

(i) start from $s^0(x) = s_o$ when seeking the maximal solution, and $s^0(x) = 0$ for the minimal solution.

(ii) define $s^{m+1}(x)$ from $s^m(x)$ by:

$$\text{(3.17)}\quad \begin{cases} -\dfrac{d^2}{dx^2}\, s^{m+1}(x) + \sigma \dfrac{1}{1 + s^m(x) + k(s^m(x))^2}\, s^{m+1}(x) = 0, \quad 0 < x < 1, \\[2ex] s^{m+1}(0) = s^{m+1}(1) = s_o \end{cases}$$

both expressions being the same since, by integrating (3.16) on (0,1) one obtains:

$$\sigma \int_0^1 F(s(x))dx = \int_0^1 s''(x)dx = s'(1) - s'(0) = 2s'(1).$$

In Figure 3.6 we see two arcs, 1 and 2, corresponding respectively to high and low activity. The corresponding steady states will be designated respectively <u>diffusion-controled</u> and <u>reaction-controled</u>.

These two types of steady states appear clearly in Figure 3.5 where the quantity $s(x)/s_o$ has been represented for several values of s_o. We distinguish "low profiles" 1,2,3, and "high profiles" 4,5,6, which are respectively diffusion-controled and reaction-controled. More precisely when s_o increases from 0 to $s_o^{**} \simeq 68$ we have "low profiles", for example 1,2,3. Then at $s_o = s_o^{**}$ there is a jump from the low profile 3 to the high profile 3.

After, the larger the value of s_o, the higher the profile. Now, if s_o decreases, the profile varies continuously through profiles 4,5,6. In particular, for that value $s_o = s_o^{**}$ for which there was a jump, the profile does not come back to the low profile 3, but remains a high profile until s_o attains the value $s_o^{*} \simeq 61$.

For this value of s_o, the profile is the high profile 6. But when s_o passes through s_o^{*}, this profile falls down to the low profile 6.

Afterwards, that is for $s_o < s_o^{*}$, when s_o continues decreasing, the steady state profiles will be the same as during the ascending phase.

3.2.4 - <u>Theoretical justification</u>: Although a more thorough analysis of the family of solutions will be performed in Section 3.4, we give here an elementary proof of the multiplicity of solutions for (3.16).

<u>Proposition 3.1</u> - For σ large enough and s_o conveniently chosen, problem (3.16) admits multiple solutions.

<u>Proof</u>: Let G be the indefinite integral:

$$G(s) = \int_0^s F(u)du, \qquad (G' = F).$$

Any solution of (3.16) must satisfy the first integral obtained by multiplying in (3.16) by $2s'(x)$:

$$-s'^2(x) + 2\sigma G(s(x)) = C, \; 0 \leq x \leq 1,$$

where C is a constant depending upon s_o and the solution.

Since $s''(x) > 0$, s is a convex function of x on [0,1], and, since $s(0) = s(1) = s_o$, we have $s'^2(0) = s'^2(1) = 2\sigma G(s_o) - C$, whence $s'(1) = -s'(0) = m$. As the right (resp. left) half part of the s concentration profile is obtained by solving the ordinary differential equation $-s''(x) + \sigma F(s(x)) = 0$ on the interval [1/2,1] (resp.) [0,1/2]) with Cauchy "initial" conditions $s(1) = s_o, s'(1) = m$ (resp. $s(0) = s_o$, $s'(0) = -m$), the s profile is symmetric with respect to $x = 1/2$. Moreover, if we restrict our considerations to the interval [1/2,1], we have $s'(x) \geq 0$, whence:

$$\frac{ds}{dx} = (2\sigma)^{1/2}(G(s) - G(\mu))^{1/2}$$

where $\mu = s(1/2)$, or

$$(3.18) \quad (2\sigma)^{1/2}(x - \tfrac{1}{2}) = \int_{\mu}^{s(x)} \frac{d\xi}{(G(\xi) - G(\mu))^{1/2}} .$$

The solution of (3.16) is equivalent to the solution of (3.18) together with the condition requiring $s(1) = s_o$:

$$(3.19) \quad \left(\frac{\sigma}{2}\right)^{1/2} = \int_{\mu}^{s_o} \frac{d\xi}{(G(\xi) - G(\mu))^{1/2}}$$

In fact (3.16) has as many solutions as we can find several values of μ satisfying (3.19), or:

$$(3.20) \quad f(\mu,s_o) = (\sigma/2)^{1/2}$$

where:

$$f(\mu,s_o) = \int_{\mu}^{s_o} \frac{d\xi}{(G(\xi) - G(\mu))^{1/2}} .$$

It can be shown that for s_o large enough the representative curve of the function $\mu \to f(\mu,s_o)$ has at least two relative extrema as curve

C in Figure 3.7 so that there are at least 3 solutions to (3.20) if σ is in a suitable interval. (Fig. 3.7 - Functions $\mu \to f(\mu, s_o)$ for values of s_o similar to those in Figure 3.5 and, for a given value of σ, determination of the corresponding values of μ.)

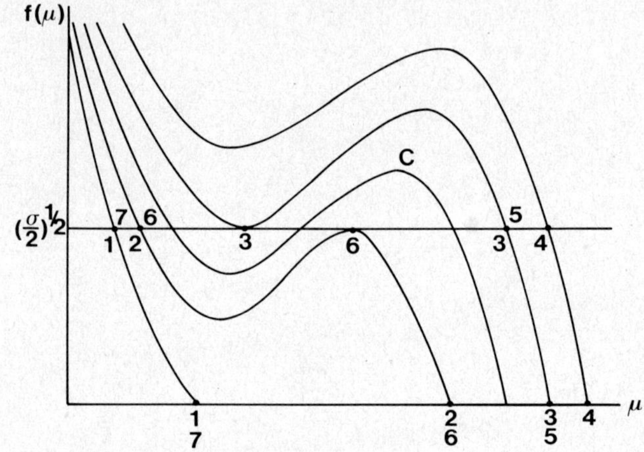

Fig. 3.7

Indeed, $f(\mu, s_o)$ can be written:

$$f(\mu, s_o) = \int_0^{s_o - \mu} \frac{d\eta}{(G(\eta+\mu) - G(\mu)^{1/2}}$$

whose derivative with respect to μ is $f'(\mu, s_o) = T_1 + T_2$ where:

$$T_1 = -1/(G(s_o) - G(\mu))^{1/2} \quad \text{and}$$

$$T_2 = (-1/2) \int_\mu^{s_o} (F(\xi) - F(\mu))/(G(\xi) - G(\mu))^{3/2} d\xi.$$

Let $\tilde{\mu}$ be such that $F'(\tilde{\mu}) < 0$ $(\tilde{\mu} > 0)$. It is easily seen that, as $s_o \to \infty$, $T_1 \to 0$ and $T_2 \to l$, $0 < l \leq \infty$, since $F(\xi) - F(\tilde{\mu}) \leq 0$ for $\xi \geq \tilde{\mu}$. Then, by choosing s_o sufficiently large $f'(\tilde{\mu}, s_o) > 0$, so that the representative curve of $f(\cdot, s_o)$ possesses an arc where f is increasing, as curve C in Figure 3.7. On the other hand it is easily seen that $f(\mu, s_o) \to \infty$ as $\mu \to 0+$ and $f(\mu, s_o) \to 0$ as $\mu \to s_o$. Thus for such a value s_o, equation (3.20) admits at least 3 solutions for σ suitably chosen.

Finally, it is easy to see that once we have found a pair $(\tilde{\sigma},\tilde{s}_o)$ such that $\tilde{\sigma} = f(\tilde{\mu},\tilde{s}_o)$ and $f'(\tilde{\mu},\tilde{s}_o) < 0$, then for every $\sigma > \tilde{\sigma}$ it is possible to find $s_o > \tilde{s}_o$ such that $\sigma = f(\tilde{\mu},s_o)$ and $f'(\tilde{\mu},s_o) < 0$.

The above proof is interesting not only because it is not always obvious whether the steady state is unique or not (see for example the glucose pump in Chapter 2), but also for the reason that it introduces the family of curves $\mu \to f(\mu,s_o)$ as s_o varies, depicted in Figure 3.7, and which is useful for explaining the profiles described in Figure 3.5.

The curves $\mu \to f(\mu,s_o)$ can be calculated for each value of s_o, and, although we have not demonstrated mathematically that there are at most three solutions for (3.20), there is some numerical evidence for this assertion to be true.

Thus if, in Figure 3.7, we follow by continuity the point common to the straight line $\mu \to (\sigma/2)^{1/2}$ and to the curve $\mu \to f(\mu,s_o)$ as s_o varies, the abscissa of this point being $\mu = s(1/2)$, we find the sequence of points numbered from 1 to 7, corresponding to the profiles in Figure 3.5 with the same numerical order. In particular we see that the two jumps 3-low → 3-high and 6-high → 6-low occur when the curve is tangent to the straight line.

Alternatively we can represent for each value of s_o the corresponding values of μ and obtain the S-shaped curve of Figure 3.8, where the order of the steady states is the same as in Figures 3.5 and 3.7. (Figure 3.8 - Representation of $\mu(=s(1/2))$ as a function of $s_o(=s(0)=s(1))$. Solid (resp. dashed) curves are used for stable (resp. unstable) steady-states. The order of the steady states is the same as in Figures 3.5 and 3.7).

Hysteresis may be observed for a given boundary value s_o: the system can be in one of two possible stable steady states depending on its past history. Thus we have a simple model of a system with memory.

The dashed curve did not appear in the numerical experiments of §3.2.3. However, we see in Figure 3.7 that whenever two distinct stable steady states exist, there is a third steady state lying between them. It will be the object of the next sections to investigate further the structure of the S-shaped curve of Figure 3.8, numerically in Section 3.3, mathematically in Sections 3.4 and 3.5.

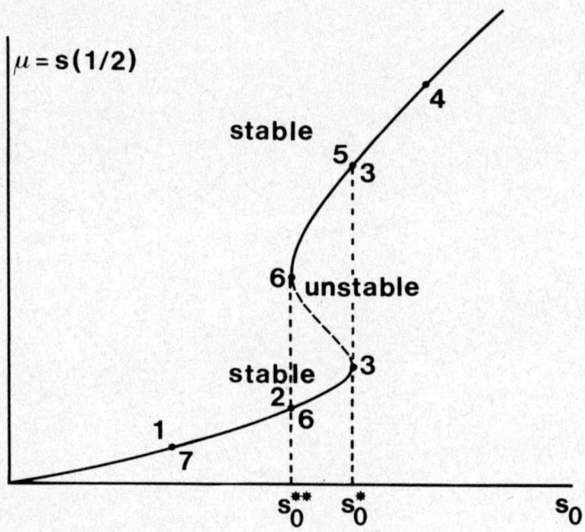

Fig. 3.8

These studies will confirm the conclusion of the present section, namely that the curve of solutions is composed of three arcs: a first one, comprising the minimal solutions for $0 \leq s_o \leq s_o^*$, which are diffusion controled, a second one, comprising the maximal solutions for $s_o < s_o^{**}$, which are reaction controled, and a third one, comprising the unstable intermediate solutions between the two turning points.

3.3 Kubicek Method of Continuation

A finite difference discretization of (3.16) gives a system of n non-linear equations:

$$(3.21) \quad \begin{cases} 2s_1 - s_2 + \rho F(s_1) = s_o \\ -s_1 + 2s_2 - s_3 + \rho F(s_2) = 0 \\ \ldots\ldots \\ -s_{i-1} + 2s_i - s_{i+1} + \rho F(s_i) = 0 \\ \ldots\ldots \\ -s_{N-1} + 2s_N + \rho F(s_N) = s_o, \end{cases}$$

where s_i is an approximation of $s(ih)$, h being the space step $h = 1/(N+1)$, and $\rho = \sigma h^2$.

If u denotes the vector of concentrations $u = [s_1, s_2, \ldots, s_N]^T$ and $\lambda = s_o$, problem (3.21) is of the form:

$$(3.22) \quad f(\lambda, u) = 0,$$

where $f : R^+ \times R^N \to R^N$ is a continuously differentiable mapping. More precisely, u stands for $[u_1, \ldots, u_N]^T$ and $f(\lambda, u)$ for $[f_1(\lambda, u_1, \ldots, u_N), \ldots, f_N(\lambda, u_1, \ldots, u_N)]^T$. The derivatives $f_\lambda(\lambda, u)$ and $f_u(\lambda, u)$ are respectively the vector:

$$f_\lambda(\lambda, u) = \left[\frac{\partial f_1}{\partial \lambda}(\lambda, u), \ \ldots, \ \frac{\partial f_N}{\partial \lambda}(\lambda, u) \right]^T$$

and the matrix:

$$f_u(\lambda, u) = \left(\frac{\partial f_i}{\partial u_j}(\lambda, u_1, \ \ldots, \ u_N) \right)_{1 \leq i, j \leq N} .$$

These equations involve the state of the system u, which is a vector of concentrations, and a parameter λ, describing the environment. We are interested in determining how the state of the system varies as this parameter is altered. In general, there may be multiple solutions to the equations, and the number and stability of these solutions undergo changes at particular values of the parameter. We desire to comprehend the interrelation among multiple solutions which have been found in Section 3.2.

The numerical method of continuation we are going to present here is general and can be applied to any problem of the form (3.22) where f is a continuously differentiable mapping of $R \times R^N$ into R^N. Stability is relative to the evolution problem $du/dt + f(\lambda, u) = 0$. The special points where the number and stability of solutions may change are of two types, turning points and bifurcation points.

In this section we shall deal only with continuation of curves of solutions with eventual turning points, the numerical exploration of bifurcating branches of solutions being treated in the next chapter. There is a well developed theory which permits the qualitative description of the behavior of solutions in the vicinity of a turning point [2]. The results of this theory will be given in the next two sections.

The Kubicek method [1] consists of parameterizing λ and u by the arc length s along the curve of solutions.

Let $(\lambda'(s), u'(s))$ be the tangent vector, where:

$$\lambda'(s) = \frac{d\lambda(s)}{ds} \quad \text{and} \quad u'(s) = \left[\frac{du_1(s)}{ds}, \ldots, \frac{du_N(s)}{ds} \right]^T .$$

Along an arc the tangent vector varies continuously, and satisfies the condition:

$$(3.23) \quad f_\lambda(s)\, \lambda'(s) + f_u(s)\, u'(s) = 0.$$

where: $f_\lambda(s) = f_\lambda(\lambda(s), u(s))$ and $f_u(s) = f_u(\lambda(s), u(s))$, and which follows from (3.22). The definition of arc length implies that the following normalization must be attached to (3.23):

$$(3.24) \quad \lambda'^2(s) + |u'(s)|^2 = 1.$$

Here $|\cdot|$ denotes the euclidean norm in R^N.

If (λ_o, u_o) is some initial point on the curve of solutions, corresponding to, say, $s = s_o$, then solving (3.22) is equivalent to finding the solution of the system of ordinary differential equations:

$$(3.25) \quad \begin{cases} \lambda' = \lambda'(s), & \lambda(s_o) = \lambda_o, \\ u' = u'(s), & u(s_o) = u_o, \end{cases}$$

where $\lambda'(s)$ and $u'(s)$ satisfy (3.23) and (3.24). In order to obtain $(\lambda'(s), u'(s))$ at the point $(\lambda(s), u(s))$ there are two cases:

First if $f_u^{-1}(s)$ exists, we solve:

$$f_u(s) y = -f_\lambda(s).$$

Since we know that: $u'(s) = \lambda'(s) y$, $\lambda'(s)$ is determined by:

(3.26) $(1 + |y|^2)\lambda'^2(s) = 1.$

There are two solutions to (3.26), which differ only according to the direction in which they trace the curve. We choose an initial direction at (λ_o, u_o) by providing a vector of signs for $\lambda'(0)$ and $u_i'(0)$, $1 \leq i \leq N$, and this direction is maintained until a turning point is encountered.

Points where $f_u^{-1}(s)$ does not exist are of course isolated along the curve of solutions, but near these points the algorithm described above loses its accuracy, due to the ill-conditioning of $f_u(s)$. The role played by λ in the above treatment may be assumed by some other component u_i $(1 \leq i \leq N)$, permitting the restatement of (3.21) which is less ill-conditioned. A strategy of full pivoting, applied to the N x (N+1) matrix $[f_\lambda \vdots f_u]$ selects the new independent variable most effectively. The details as well as a Fortran code may be found in Kubicek [1].

Having obtained $(\lambda'(s), u'(s))$ we can proceed to integrate (3.25) numerically, using the Adams-Bashforth formulas of maximal order 4, again following Kubicek. The step length and order of the integration formula are adjusted automatically. In particular, whenever the independent variable is changed, the order of the integration formula is reduced to 1, and the step size is reduced by an empirically chosen factor (10). The numerical integration is a predictor step, providing an estimate $(\hat{\lambda}(s+h), \hat{u}(s+h))$ of the next point on the branch, $(\lambda(s+h), u(s+h))$, see Figure 3.9, (Scheme for tracing a curve of solutions in $R \times R^N$ from known point $(\lambda(s), u(s))$ by predictor step $(\lambda(s), u(s)) \to (\hat{\lambda}(s+h), \hat{u}(s+h))$, followed by corrector step $(\hat{\lambda}(s+h), \hat{u}(s+h)) \to (\lambda(s+h), u(s+h))$.

This estimate is not without error, so it is taken as a starting point for a Newton iteration which refines the estimate up to some error tolerance. In case the independent variable is λ, for example, the Newton method for variable u is:

$$\begin{cases} f_u(\hat{\lambda}(s+h), u^{(k)})(u^{k+1} - u^k) = -f(\hat{\lambda}(s+h), u^k) \\ u^o = \hat{u}(s+h) \end{cases}$$

We use the number of Newton corrections as a basis for a second stage of control of step size in the predictor code. Whenever three or fewer iterations suffice to satisfy the error tolerance, the step size is increased by an empirically chosen factor (1.5), up to some maximum step length, selected so as to produce curves of adequate density.

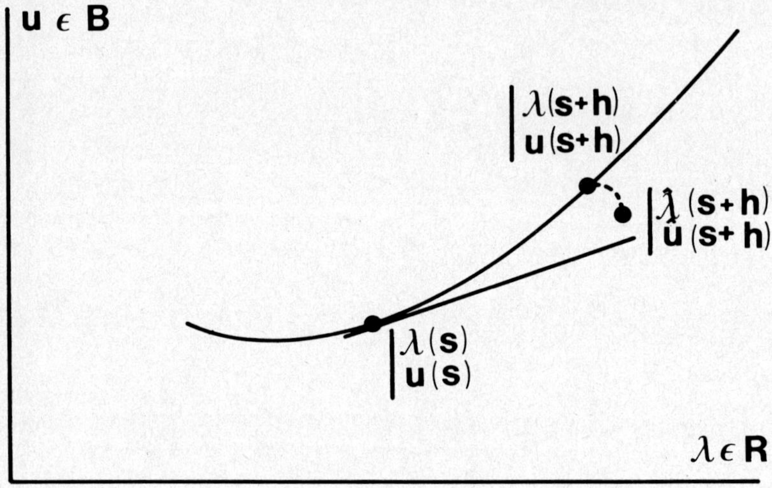

Fig. 3.9

The application of the Kubicek method to problem (3.21) with N = 20 gives the results depicted in Figure 3.10. Evidently, (0;0,...,0) is a solution to $f(\lambda,u) = 0$ and we take it as a starting point on the curve. Only slightly less evident:

$$\frac{\partial u}{\partial \lambda} / (0;0,\dots,0) > 0,$$

so the initial direction vector is obtained = (+1;+1,...,+1). The resulting collection of solution components is illustrated in Figure 3.10 where only the components corresponding to i = 1, 3, and 10 have been shown. Linearized stability was inferred from the change in sign of the determinant of f_u. There are two turning points on the curve, for $s_o = s_o^*$ and $s_o = s_o^{**}$, at which this determinant changes sign. Since the determinant of f_u^o does not change sign between the turning points, we infer that the steady state is unstable all along the intermediate arc lying between these two turning points.

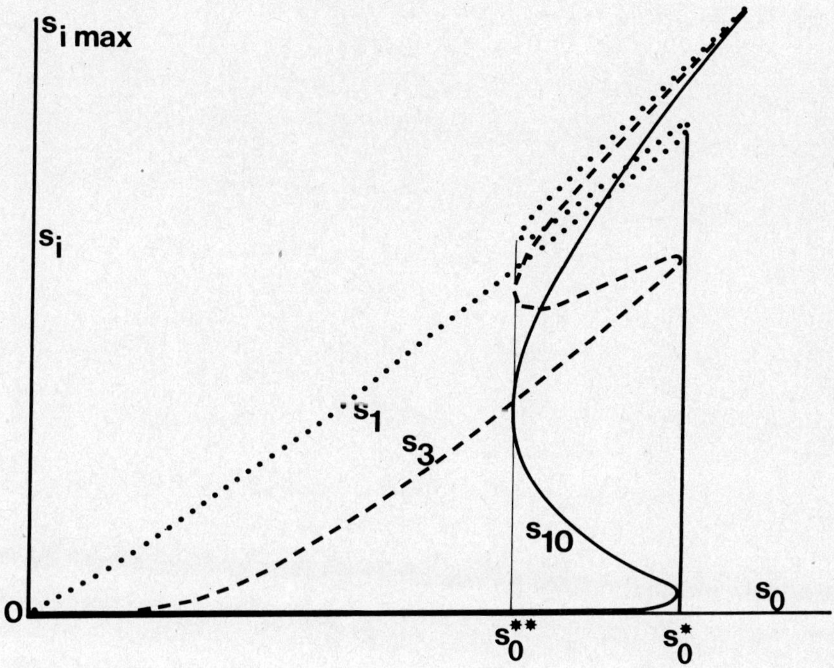

Fig. 3.10

(Fig. 3.10 - Components s_i of the solution to (3.21) for i = 1,3, and 10. The abscissa is s_o, the ordinate is s_i. $s_{i\,max}$ = 30.60 for i = 1, 28.92 for i = 3, and 26.26 for i = 10. $s_o^* = 21.5$, $s_o^{**} = 28.9$. $\sigma = 1200$, $h = 1/20$, $\rho = 3$, and $k = 1$).

3.4 Mathematical Analysis of the S-shaped Curve

We rewrite the problem (3.16) by calling λ the parameter s_o, and $s(x,\lambda)$ the solution of:

$$(3.27) \quad \begin{cases} -s''(x,\lambda) + \sigma F(s(x,\lambda)) = 0, & 0 < x < 1, \\ s(0,\lambda) = s(1,\lambda) = \lambda, \end{cases}$$

where the primes denote derivation with respect to x.

Let us think of s as the mapping $\lambda \to s(\cdot,\lambda)$ where $s(\cdot,\lambda)$ is the function $x \to s(x,\lambda)$. We then write $s(\lambda)$ rather than $s(\cdot,\lambda)$, corresponding to $s(x,\lambda)$ via $s(\lambda)(x) = s(x,\lambda)$.

Let σ be such that problem (3.27) admits, for a range of conveniently chosen boundary values λ, multiple solutions. We have shown, in the proof of Proposition 3.1, how to find such a σ. Our aim in this section is to clarify the structure of the family of solutions of equations (3.27) when σ is held fixed and λ varies. It seems likely, from the discussion of Section 3.2 and the numerical results of section 3.3, that the points $(\lambda, s(\lambda))$ lie on an S-shaped curve. Indeed it is possible to demonstrate precise statements about the arcs of maximal and minimal solutions, and in particular to prove that they are continuous curves ending at turning points. Our proofs follow arguments of Crandall and Rabinowitz [2,5,6], Brauner and Nicolaenko [7], and Mignot and Puel [8]. We shall first see what can be done by using upper and lower solution techniques. Then we shall add to our tool kit the Implicit Function Theorem, which is very useful for proving existence and uniqueness of a continuous curve of solutions passing through a point. Lastly we shall investigate the behavior of the curve of solutions near a turning point.

In the following $C([0,1])$ will denote the Banach space of continuous functions $u : [0,1] \to R$ equipped with the norm:

$$\|u\|_o = \max_{x \in [0,1]} |u(x)|,$$

and $C^k([0,1])$ the Banach space of k-times continuously differentiable functions $u : [0,1] \to R$ under the norm:

$$\|u\|_k = \max_{x \in [0,1]} (|u(x)| + |u'(x)| + \ldots + |u^{(k)}(x)|).$$

It is easy to demonstrate the following result for problem (3.27):

Proposition 3.2 -

(i) For every $\lambda \geq 0$, there exists a maximal solution $\hat{s}(\cdot,\lambda)$ and a minimal solution $\bar{s}(\cdot,\lambda)$, satisfying:

$$0 \leq \bar{s}(x,\lambda) \leq \hat{s}(x,\lambda) \leq \lambda, \qquad 0 \leq x \leq 1.$$

(ii) The mappings $\lambda \to \bar{s}(\cdot,\lambda)$ and $\lambda \to \hat{s}(\cdot,\lambda)$ are monotone increasing on $[0,+\infty[$.

(iii) For every $\lambda \geq 0$,

(3.28) $\overline{s}(x,\lambda) \geq \lambda[ch(\sigma^{1/2}(x - 1/2))]/ch(\sigma^{1/2}/2) \geq \lambda/ch(\sigma^{1/2}/2)$.

Proof -

(i) It is obvious that 0 is a lower solution and λ an upper solution. Hence we can apply Proposition 1.2. Alternatively we can apply Theorem 1.1 within the following framework.

E is the Banach space $C([0,1])$ of continuous functions on $[0,1]$, endowed with the norm: $||u||_o = \max_{0\leq x\leq 1}|u(x)|$. E is an ordered Banach space for the natural ordering: $u_1 \leq u_2 \Leftrightarrow u_1(x) \leq u_2(x)$ for every $x \in [0,1]$. We consider the order interval $[\overline{y},\hat{y}]$ where $\overline{y}(x) \equiv 0$ and $\hat{y}(x) \equiv \lambda$, and define $f: [\overline{y},\hat{y}] \to E$ by: $u = f(v)$ if:

$$-u'' + \sigma uG(v) = 0, \qquad u(0) = u(1) = \lambda,$$

where:

(3.29) $G(v) = 1/(1 + v + kv^2)$.

We verify that f is a monotone increasing mapping. If $v_1 \leq v_2$ and $w = u_2 - u_1$, we have, since G is monotone decreasing on R_+,

$$-w'' + \sigma wG(v_2) = \sigma u_1(G(v_1) - G(v_2)) \geq 0,$$

which, together with $w(0) = w(1) = 0$, implies $w \geq 0$. We verify that $\overline{y} \leq f(\overline{y})$ and $f(\hat{y}) \leq \hat{y}$. $u = f(\overline{y})$ is defined by:

(3.30) $-u'' + \sigma u = 0, \quad u(0) = u(1) = \lambda$.

The solution of (3.30) is:

(3.31) $u = \lambda[ch(\sigma^{1/2}(x - 1/2)]/ch(\sigma^{1/2}/2)$

which at evidence satisfies $u \geq 0$. $u = f(\hat{y})$ is defined by: $-u'' + \sigma uG(\lambda) = 0$, $u(0) = u(1) = \lambda$. Let $w = \lambda - u$. We have $-w'' +$

$\sigma wG(\lambda) = \sigma F(\lambda)$, $w(0) = w(1) = 0$.

From the Maximum Principle $w \geq 0$, hence $u \leq \lambda$. Lastly we verify that <u>f is compact</u>. Let (v_n) be a sequence in $[\overline{y},\hat{y}]$. The definition of u_n, $-u_n'' + \sigma u_n G(v_n) = 0$, $u_n(0) = u_n(1) = \lambda$, implies that, for every $x \in [0,1]$, $0 \leq u_n(x) \leq \lambda$, $0 \leq u_n''(x) \leq \sigma\lambda$, $|u_n'(x)| \leq \sigma\lambda$, the last inequality resulting from the fact that there is an $x_o \in [0,1]$ such that $u_n'(x_o) = 0$, and, for every $x \in [0,1]$,

$$u_n'(x) = \int_{x_o}^{x} u_n''(\xi)d\xi.$$

The u_n remain in a bounded set of $C^2([0,1])$. The compactness of f follows from the compactness of the natural embedding of $C^2([0,1])$ into $C([0,1])$.

(ii) Let $\lambda_1 < \lambda_2$. $\overline{s}(\cdot,\lambda_2)$ is an upper solution for problem (3.27) with $\lambda = \lambda_1$. Since the minimal solution of this problem either in interval $[0,\lambda_1]$ or in interval $[0,\overline{s}(\cdot,\lambda_2)]$ is the limit of $f^k(0)$ as $k \to \infty$, the minimal solution is the same in both intervals, so that $\overline{s}(\cdot,\lambda_1) \leq \overline{s}(\cdot,\lambda_2)$. Similarly $\hat{s}(\cdot,\lambda_1)$ is a lower solution for problem (3.27) with $\lambda = \lambda_2$, whose maximal solution is the same either in $[0,\lambda_2]$ or in $[\hat{s}(\cdot,\lambda_1),\lambda_2]$, implying $\hat{s}(\cdot,\lambda_1) \leq \hat{s}(\cdot,\lambda_2)$.

(iii) results from the fact that, according to the proof of Theorem 1.1, $0 = \overline{y} \leq f(\overline{y}) = u \leq \ldots \leq \overline{s}(\cdot,\lambda)$, u being given by (3.31).

$\overline{s}(\lambda)$ and $\hat{s}(\lambda)$ may be the same. This is the case in particular when λ is close to 0 or sufficiently large.

<u>Proposition 3.3</u> - The solution of problem (3.27) is unique in the following cases:

(i) $\lambda \leq k^{-1/2}$,

(ii) $\lambda > \pi^{-1}\sigma^{1/2}k^{-1/2}ch(\sigma^{1/2}/2)$.

<u>Proof</u> - Let s_1 and s_2 be two solutions of (3.27), and $w = s_1 - s_2$.

We have:

$$-w'' + qw = 0, \qquad w(0) = w(1) = 0$$

where:

$$q = \sigma(1 - ks_1s_2)/[(1 + s_1 + ks_1^2)(1 + s_2 + ks_2^2)]$$

In the case (i), since $0 < s_1, s_2 < \lambda < k^{-1/2}$, we have $q \geq 0$, hence $w = 0$.

On the other hand, $q > -\sigma/(ks_1s_2)$ and, from (3.28), $s_1, s_2 \geq \lambda/ch(\sigma^{1/2}/2$ Thus $q > -\sigma ch^2(\sigma^{1/2}/2)/(k\lambda^2)$. In the case (ii), $q > -\pi^2$, that is $q \geq -\pi^2 + \alpha$ for some $\alpha > 0$, so that:

$$0 = \int_o^1 (-w'' + qw)w\,dx \geq \int_o^1 w'^2dx + (-\pi^2 + \alpha)\int_o^1 w^2dx$$

$$> \alpha \int_o^1 w^2dx$$

(By applying Poincaré Inequality $\int_o^1 w'^2dx \geq \pi^2 \int_o^1 w^2dx$). Thus $w = 0$.

The families of minimal and maximal solutions can be a bit more precise.

Proposition 3.4 - The mapping $\lambda \to \bar{s}(\cdot,\lambda)$ (resp. $\lambda \to \hat{s}(\cdot,\lambda)$) is left (resp. right) - continuous from $[0,+\infty[$ into $C^3([0,1])$.

Proof - Let λ_o be any fixed value of $\lambda(\lambda_o > 0)$, and consider the family of functions $\bar{s}(\cdot,\lambda)$ indexed by λ with $0 \leq \lambda \leq \lambda_o$. Since this family is monotone increasing and bounded above by $\bar{s}(\cdot,\lambda_o)$, as $\lambda \uparrow \lambda_o$, the $\bar{s}(\cdot,\lambda)$ converge pointwise and in $L^2(0,1)$ towards some limit $v \leq \bar{s}(\cdot,\lambda_o)$. It follows that $F(\bar{s}(\cdot,\lambda)) \to F(v)$ in $L^2(0,1)$. Let $u(\lambda) = u(\cdot,\lambda)$ be the function $u(\cdot,\lambda) = \bar{s}(\cdot,\lambda) - \lambda$, satisfying $-u''(x,\lambda) = -\sigma F(\bar{s}(x,\lambda))$, $0 < x < 1$, $u(0,\lambda) = u(1,\lambda) = 0$, and w the solution of $-w'' = -F(v)$, $w(0) = w(1) = 0$. Since $F(\bar{s}(\cdot,\lambda)) \to F(v)$ in $L^2(0,1)$, $u(\cdot,\lambda) \to w$ in $H^2(0,1) \cap H_o^1(0,1)$. But $u(\cdot,\lambda) = \bar{s}(\cdot,\lambda) - \lambda \to v - \lambda_o$ in $L^2(0,1)$. Hence $w = v - \lambda_o$ and the limit v of $\bar{s}(\cdot,\lambda)$ as $\lambda \uparrow \lambda_o$ satisfies:

(3.22) $\quad -v'' + \sigma F(v) = 0, \qquad v(0) = v(1) = \lambda_o.$

We already know that $v \leq \bar{s}(\cdot,\lambda_o)$. On the other hand, since v satisfies (3.22) and belongs to the order interval $[0,\lambda_o]$, we have $\bar{s}(\cdot,\lambda_o) \leq v$. Therefore $v = \bar{s}(\cdot,\lambda_o)$. Lastly, since $H^2(0,1)$ is continuously embedded in $C^1([0,1])$, we also have $\bar{s}(\cdot,\lambda) \to \bar{s}(\cdot,\lambda_o)$ in $C^1([0,1])$, $F(\bar{s}(\cdot,\lambda)) \to F(\bar{s}(\cdot,\lambda_o))$ in $C^1([0,1])$, whence $\bar{s}(\cdot,\lambda) \to \bar{s}(\cdot,\lambda_o)$ in $C^3([0,1])$.

The right continuity proof for $\lambda \to s(\cdot,\lambda)$ would be similar.

Proposition 3.4 is in agreement with Figures 3.8 and 3.10 where it can be seen, in particular, that, as $s_o \to s_o^*$ from below, $\bar{s}(\cdot,s_o) \to \bar{s}(\cdot,s_o^*)$, whereas as $s_o \to s_o^*$ from above, $\bar{s}(\cdot,s_o)$ does not converge towards $\bar{s}(\cdot,s_o^*)$. Similar remarks can be made for $s_o = s_o^{**}$.

In order to obtain further information about the curve of solutions, we need to employ the Implicit Function Theorem. For the sake of completeness we give below a version of this theorem which was stated by Crandall and Rabinowitz in [5], and for the proof of which they refer to [9].

Theorem 3.1 (The Implicit Function Theorem) - Let X, Y, Z be three Banach spaces, f a continuous mapping of an open subset A of X x Y into Z. Let the map $y \to f(x,y)$ of $A_x = \{y \in Y : (x,y) \in A\}$ into Z be differentiable in A_x for each $x \in X$ such that $A_x \neq \emptyset$, and assume the derivative of this map (denoted by f_y) is continuous on A. Let $(x_o,y_o) \in A$ be such that $f(x_o,y_o) = 0$, and $f_y(x_o,y_o)$ is a linear homeomorphism of Y onto Z. Then there are neighborhoods U of x_o in X and V of y_o in Y such that:

(i) $U \times V \subset A$.

(ii) There is exactly one function $u:U \to V$ satisfying $f(x,u(x)) = 0$ for $x \in U$.

(iii) The mapping u of (ii) is continuous.

If, moreover, the mapping f is k-times continuously differentiable on A, then (iii) above may be replaced by:

(iv) u is k-times continuously differentiable.

Due to its amazing scope and simplicity, the implicit function theorem is a very useful analytical tool. A nice introduction to the material involved in this theorem and its applications can be found in [10].

We begin with an application of the implicit function theorem to the following situation: $X = R$, $Y = C_0^2([0,1])$, Banach space of twice continuously differentiable functions on $[0,1]$ vanishing at $x = 0$ and $x = 1$, equipped with the norm $||u||_2$, $Z = C([0,1])$, equipped with the norm $||u||_0$, $A = X \times Y$, $f : (\lambda,u) \to -(u + \lambda)'' + \sigma F(u + \lambda)$, $x_0 = 0$ and $y_0 = 0$. It is easily seen that $f(x_0,y_0) = 0$ and that $f_y(x_0,y_0)$: $w \to -w'' + \sigma w$ is an homeomorphism of Y onto Z. Then there are neighborhoods $U =]-\varepsilon,+\varepsilon[$ of 0 in R and V of 0 in $C_0^2([0,1])$ such that:

(i) there is exactly one function $u:U \to V$ satisfying $f(\lambda,u(\lambda)) = 0$ for $\lambda \in U$.

(ii) the mapping u of (i) is continuous.

Moreover, since $f_u(0,0)$ is non singular, $f_u(\lambda,u(\cdot,\lambda))$ is non singular for $0 \leq \lambda < \varepsilon$ if ε is chosen sufficiently small.

Thus we have the:

<u>Proposition 3.5</u> - There is an interval $[0,\varepsilon[$ of R and a neighborhood V of 0 in $C_0^2([0,1])$ such that:

(i) there is exactly one function $u:[0,\varepsilon[\to V$ satisfying:

(3.33) $-u''(\cdot,\lambda) + \sigma F(u(\cdot,\lambda) + \lambda) = 0$

for $\lambda \in [0,\varepsilon[$. Moreover:

(ii) the mapping u of (i) is continuous,

(iii) the operator $v \to -v'' + \sigma F'(u(\cdot,\lambda) + \lambda)v$ is invertible as a map from $C_0^2([0,1])$ into $C([0,1])$.

<u>Proposition 3.6</u> - There is a number $\bar{\lambda}$, $0 < \bar{\lambda} \leq \infty$, which is maximal with respect to the following properties:

(i) existence of a continuous function $u:[0,\bar{\lambda}[\to C_o^2([0,1])$ for which:

(3.34) $-u''(\cdot,\lambda) + \sigma F(u(\cdot,\lambda) + \lambda) = 0$, and:

(ii) for every $\lambda \in [0,\bar{\lambda}[$, the operator $v \to -v'' + \sigma F'(u(\cdot,\lambda) + \lambda)v$ is invertible as a map from $C_o^2([0,1])$ into $C([0,1])$.

Moreover, $\bar{\lambda}$ and $u(\lambda)$ are unique.

<u>Proof</u> - We are interested by those intervals $[0,a[$ on which we can define a continuous function $u:[0,a[\to C_o^2([0,1])$ satisfying, for every $\lambda \in [0,a[$,

(i) $f(\lambda,u(\lambda)) = 0$ and

(ii) $f_u(\lambda,u(\lambda))$ is non singular.

Of course, for λ small enough, such a function coincides with the function u provided by Proposition 3.5, and, in particular the curve $(\lambda,u(\lambda))$ emanates for the point (0,0).

Moreover, if $u_1:[0,a_1[\to C_o^2([0,1])$ and $u_2:[0,a_2[\to C_o^2([0,1])$ are two functions of λ satisfying (i) and (ii) with $a_1 \leq a_2$, then $u_1(\lambda) = u_2(\lambda)$ for every $\lambda \in [0,a_1[$. If not, since u_1 and u_2 are continuous functions of λ, there exists a closed interval $[0,b]$ with $b < a_1$, such that $u_1(\lambda) = u_2(\lambda)$ for every $\lambda \in [0,b]$ but, whatever small η, and whatever the neighborhood V of $u_1(b)$ in $C_o^2([0,1])$, there exists at least one $\lambda \in [b,b+\eta[$ for which $u_1(\lambda) \neq u_2(\lambda)$ and $u_1(\lambda) \in V$, $u_2(\lambda) \in V$. Since the implicit function theorem, applied at $x_o = b$, $y_o = u_1(b)$, implies the existence of neighborhoods $]b-\eta,b+\eta[$ of b and V of $u_1(b)$ such that there is an unique solution of $f(\lambda,u) = 0$ satisfying $(\lambda,u) \in]b-\eta,b+\eta[\times V$, there is a contradiction.

Now it suffices to define $\bar{\lambda}$ as the least upper bound of the numbers a such that on interval $[0,a[$ there exists a function $u:[0,a[\to C_o^2([0,1])$ satisfying (i) and (ii). It follows that $\bar{\lambda}$ and the function $u:[0,\bar{\lambda}[\to C_o([0,1])$ satisfying (i) and (ii) are uniquely defined.

Let $s(\lambda) = s(\cdot,\lambda) = u(\cdot,\lambda) + \lambda$ where $u(\cdot,\lambda)$ is the function provided by Proposition 3.6, and $\bar{u}(\lambda) = \bar{u}(\cdot,\lambda) = \bar{s}(\lambda) - \lambda$ where $\bar{s}(\lambda)$ is the minimal solution of problem (3.27). Before proving that $s(\lambda) = \bar{s}(\lambda)$ for $\lambda \in [0,\bar{\lambda}[$, we need the following:

<u>Lemma 3.1</u> - Suppose that $\lambda_o \in [0,\bar{\lambda}[$ and satisfies $s(\lambda_o) = \bar{s}(\lambda_o)$. Then there exists $\eta > 0$ such that $s(\lambda) = \bar{s}(\lambda)$ for every $\lambda \in [\lambda_o, \lambda_o + \eta[$.

<u>Proof</u> - Since, for $\lambda > \lambda_o, s(\lambda_o) = \bar{s}(\lambda_o) \leq \bar{s}(\lambda) \leq s(\lambda)$, we have successively:

$$\|\bar{s}(\lambda) - s(\lambda_o)\|_o \leq \|s(\lambda) - s(\lambda_o)\|_o \to 0 \text{ as } \lambda \downarrow \lambda_o,$$

$$\|\bar{u}(\lambda) \quad u(\lambda_o)\|_o = \|\bar{s}(\lambda) - s(\lambda_o) - (\lambda-\lambda_o)\|_o \to 0,$$

$$(\bar{u}(\lambda) - u(\lambda_o))'' = \sigma(F(\bar{s}(\lambda)) - F(s(\lambda_o))) \to 0 \quad \text{in} \quad C([0,1]),$$

$$\bar{u}(\lambda) \to u(\lambda_o) \quad \text{in } C_o^2([0,1]).$$

By the implicit function theorem there are neighborhoods $]\lambda_o-\eta,\lambda_o+\eta[$ of λ_o in R and V of $u(\lambda_o)$ in $C_o^2([0,1])$ such that there is exactly one function $u\colon]\lambda_o-\eta,\lambda_o+\eta[\to V$ satisfying $f(\lambda,u(\lambda)) = 0$ for $\lambda \in]\lambda_o-\eta,\lambda_o+\eta[$. Moreover, since η can be chosen so that $\bar{u}(\lambda) \in V$ for $\lambda \in [\lambda_o,\lambda_o+\eta[$, it follows that $u(\lambda) = \bar{u}(\lambda)$ for $\lambda \in [\lambda_o,\lambda_o+\eta[$. This completes the proof of the Lemma.

<u>Proposition 3.7</u> - $s(\lambda) = \bar{s}(\lambda)$ for $\lambda \in [0,\bar{\lambda}[$. Moreover if $\bar{\lambda} < \infty$ then $s(\bar{\lambda})$ is well defined by:

$$s(\bar{\lambda}) = \lim_{\lambda \uparrow \bar{\lambda}} s(\lambda) = \bar{s}(\bar{\lambda}).$$

<u>Proof</u> - Since $s(0) = \bar{s}(0)$, Lemma 3.1 implies the existence of an interval $[0,\eta[$ on which $s(\lambda) = \bar{s}(\lambda)$. Let $[0,b[$ be its maximal extension, that is let b be the least upper bound of those a such that $s(\lambda) = \bar{s}(\lambda)$ for $\lambda \in [0,a[$. Suppose $b < \bar{\lambda}$. Since $s(\lambda) = \bar{s}(\lambda)$ for $\lambda \in [0,b[$ and $\bar{s}\colon \lambda \to \bar{s}(\lambda)$ is left-continuous as a mapping of R^+ into $C^2([0,1])$, then $s(b) = \bar{s}(b)$. Thus, by Lemma 3.1, $s(\lambda) = \bar{s}(\lambda)$ on $[0,b+\varepsilon[$, $\varepsilon > 0$, which contradicts the maximality of b. Hence $b = \bar{\lambda}$. Moreover, if $\bar{\lambda} < \infty$ then $s(\bar{\lambda})$ is well defined since $s(\lambda) = \bar{s}(\lambda) \to \bar{s}(\bar{\lambda})$ as $\lambda \to \bar{\lambda}$ from below.

We now have all the elements for proving:

Proposition 3.8 - If there is at least one value λ_o of λ such that $\overline{s}(\lambda_o) \neq \hat{s}(\lambda_o)$, then $\overline{\lambda} < \infty$.

Proof - Suppose $\overline{\lambda} = +\infty$. By Proposition 3.7 $s(\lambda) = \overline{s}(\lambda)$ for every $\lambda \in [0,+\infty[$. By Proposition 3.3 $\overline{s}(\lambda) = \hat{s}(\lambda)$ for λ large enough. Let $]\lambda_1,+\infty[$ be the maximal extension of those intervals $]a,+\infty[$ on which $s(\lambda) = \hat{s}(\lambda)$. From the continuity of $\lambda \to s(\lambda)$ and the right continuity of $\lambda \to \hat{s}(\lambda)$, we have $s(\lambda_1) = \hat{s}(\lambda_1)$. Of course $\lambda_1 \geq \lambda_o$. Now consider the monotone increasing family $\{\hat{s}(\lambda) | \lambda < \lambda_1\}$, bounded above by $\hat{s}(\lambda_1)$.

This family converges pointwise and in $L^2(0,1)$ towards some element $v \leq \hat{s}(\lambda_1)$. As in the proof of Proposition 3.4, it is easy to check that $\hat{s}(\lambda)$ converges in $C^2([0,1])$ towards v, and v satisfies $-v''(x) + \sigma F(v(x)) = 0$ for $0 < x < 1$, together with $v(0) = v(1) = \lambda_1$. Thus $\overline{s}(\lambda_1) \leq v \leq \hat{s}(\lambda_1)$. But, since $\overline{s}(\lambda_1) = \hat{s}(\lambda_1)$, $v = s(\lambda_1)$. By the implicit function theorem there are neighborhoods $]\lambda_1-\eta,\lambda_1+\eta[$ of λ_1 in R and V of $s(\lambda_1)$ in $C^2([0,1])$ such that there is exactly one function $s:]\lambda_1-\eta,\lambda_1+\eta[\to V$ satisfying $f(\lambda,s(\lambda) - \lambda) = 0$ for $\lambda \in]\lambda_1-\eta,\lambda_1+\eta[$. Moreover we can choose η small enough for $\hat{s}(\lambda)$ to be in V for $\lambda \in]\lambda_1-\eta,\lambda_1]$. But this cannot occur because the interval $[\lambda_1,+\infty[$ is maximal. This completes the proof of the Proposition.

In order to obtain further information on the limit point $(\overline{\lambda},s(\overline{\lambda}))$, we shall use the two following theorems of Crandall and Rabinowitz [2], which give a description of the behavior of the curve of solutions for $f(\lambda,u) = 0$ near this limit point. As Theorem 3.2 follows at once from the implicit function theorem, we shall give the details of its proof for the sake of completeness, whereas for Theorem 3.3, which is more complicated to demonstrate, we refer to the proof given in [2].

Theorem 3.2 - Let Y and Z be Banach spaces, $(\lambda_o,u_o) \in R \times Y$ and let f be a continuously differentiable mapping of an open neighborhood of (λ_o,u_o) into Z. Let $f(\lambda_o,u_o) = 0$, $N(f_u(\lambda_o,u_o)) = \text{span}\{\phi_1\}$ be one-dimensional and codim $R(f_u(\lambda_o,u_o)) = 1$. Let $f_\lambda(\lambda_o,u_o) \notin R(f_u(\lambda_o,u_o))$. If X is a complement of span $\{\phi_1\}$ in Y, then the solutions of $f(\lambda,u) = 0$ near (λ_o,u_o) form a curve $(\lambda(s),u(s)) = (\lambda_o + \tau(s),\ u_o + s\phi_1 + x(s))$ where $s \to (\tau(s),x(s)) \in R \times X$ is a continuously differentiable function near $s = 0$ and $\tau(0) = \tau'(0) = 0$, $x(0) = x'(0) = 0$. Moreover, if f is k-times continuously differentiable (analytic), so are $\tau(s),x(s)$.

Here dim and codim are abbreviations for dimension and codimension, respectively, (the codimension of a subspace W of Z is the dimension of Z/W), and N(T),R(T) denote the null space and range of a linear mapping T.

No confusion should be made between the parameter s in the statement of Theorem 3.2 and the solution of Problem 3.27. Let:

$$f_\lambda^0 = f_\lambda(\lambda_o,u_o)$$

and

$$f_u^0 = f_u(\lambda_o,u_o).$$

Proof - Define a function g by:

$$g(s,\tau,x) = f(\lambda_o + \tau,\ u_o + s\phi_1 + x)$$

for $(s,\tau,x) \in R \times R \times X$. Observe that the partial derivatives g_τ and g_x are continuous in (s,τ,x). In addition:

$$g(0,0,0) = f(\lambda_o,u_o) = 0$$

and the assumptions of Theorem 3.2 imply that the Fréchet derivative of the map $(\tau,x) \to g(0,\tau,x)$ at $(\tau,x) = (0,0)$, which is the linear map:

$$(\hat{\tau},\hat{x}) \to f_\lambda^0\hat{\tau} + f_u^0\hat{x}$$

of R x X into Z, is an isomorphism onto Z. The implicit function theorem implies the existence of functions τ and x possessing the properties asserted in the theorem. In particular, since f is continuously differentiable, so are τ(s) and x(s). More precisely the derivatives τ'(s) and x'(s) are given by:

$$f_\lambda(s)\tau'(s) + f_u(s)x'(s) + f_u(s)\phi_1 = 0$$

where $f_\lambda(s)$ and $f_u(s)$ are the values of f_λ and f_u at $(\lambda_o + \tau(s),$ $u_o + s\phi_1 + x(s))$. For s = 0 we obtain $f_\lambda^0\tau'(0) + f_u^0x'(0) + f_u^0\phi_1 = 0$. Since $f_u^0\phi_1 = 0$, it follows that $\tau'(0) = 0$ and $x'(0) = 0$.

Theorem 3.3 - Let 0 be a simple eigenvalue of f_u^0. There are continu-

(Fig. 3.11) is a graph of $u(A)$, A being any point in Ω.

A complete treatment of point (i) can be found in [8]. Point (ii) uses arguments of Crandall and Rabinowitz [2, page 177].

3.5 Stability

In this section we study the stability of the equilibrium solutions for the one-dimensional case:

$$(3.41)\quad \begin{cases} u_t - u_{xx} + r(u) = 0, & 0 < x < 1, \quad t > 0, \\ u(0,t) = u(1,t) = 0, & t > 0. \end{cases}$$

Here $u(x,t) = s(x,t) - s_o$ and $r(u) = \sigma F(s_o + u)$, and the evolution equation (3.41) must be supplemented with the initial condition:

$$(3.42)\quad u(x,0) = u_o(x), \qquad 0 < x < 1.$$

We say v is an <u>equilibrium point</u> of (3.41) if $u(x,t) = v(x)$ is a solution of (3.41). Namely:

$$(3.43)\quad \begin{cases} -v''(x) + r(v(x)) = 0, & 0 < x < 1, \\ v(0) = v(1) = 0. \end{cases}$$

Equivalently, we say v is a <u>stationary</u> or <u>steady state</u>. Indeed, "stationary" is more appropriate than "steady", since stationary means "fixed in a position, not moving", whereas steady may mean "having a continuous movement", and may be applied, for example, to oscillations. However, when dealing with solutions of (3.43) as particular solutions of (3.41), we may and shall assume that these three expressions, equilibrium point, stationary state, and steady state have the same meaning.

We say v is a <u>stable</u> equilibrium point of (3.41) provided that we can keep the solution of (3.41)-(3.42) as close to v as we desire for all $t \geq 0$ by choosing u_o sufficiently close to v. More precisely, for any $\varepsilon > 0$, there exists $\delta > 0$ such that any solution u of (3.41) with $\|u(0) - v\| < \delta$ satisfies $\|u(t) - v\| < \varepsilon$ for all $t > 0$. Here we write $u(t)$ rather than $u(\cdot,t)$ for the function $x \to u(x,t)$, and $\|\cdot\|$ is the usual norm in $H_o^1(0,1)$.

We say v is an <u>asymptotically stable</u> solution of (3.41) if it is stable and we can also guarantee that $u(t) \to v$ in $H_o^1(0,1)$ as $t \to \infty$ by choosing $\|u(0) - v\|$ sufficiently small.

We say v is <u>unstable</u> if it is not stable. Specifically, there exist $\varepsilon_o > 0$ and $\{u_{on}, n \geq 1\}$ with $\|u_{on} - v\| \to 0$ as $n \to \infty$, but for all n, we can guarantee that there exists a time t such that the solution $u_n(t)$ of (3.41) with initial condition u_{on} satisfies:

$$\|u_n(t) - v\| \geq \varepsilon_o > 0.$$

In finite dimensions, that is $u \in R^N$ and $f:R^N \to R^N$ it is well known that the solution v of $f(u) = 0$ is an asymptotically stable equilibrium point of $du/dt + f(u) = 0$ if the eigenvalues of $f_u(v)$ (the Jacobian matrix $(\partial f_i/\partial u_j)$ evaluated at $u = v$) all have positive real parts and is unstable if $f_u(v)$ has an eigenvalue with a negative real part.

In infinite dimensions, as is the case for problem (3.41), analogues of this remain true.

More precisely, consider the nonlinear problem (3.41) where the function $r:R \to R$ is continuously differentiable, its derivative being bounded. Suppose v is an equilibrium point of (3.41), such that:

$$r(v + z) = r(v) + r'(v)z + h(z)$$

with

$$(3.44) \quad |h(z)| = O(\|z\|^2)$$

where $|\cdot|$ (resp. $\|\cdot\|$) denotes the norm in $L^2(0,1)$ (resp. $H_o^1(0,1)$).

Moreover we assume u_o to be in $H_o^1(0,1)$.

Let L be the "linearized operator" from $C_o^2([0,1])$ into $C([0,1])$ $L:w \to -w'' + r'(v)w$.

It is well known that the eigenvalue problem:

$$\begin{cases} Lw = \mu w, & 0 < x < 1, \\ w(0) = w(1) = 0 \end{cases}$$

possesses a sequence μ_i, $i = 1,2,\ldots$, of real eigenvalues converging towards $+\infty$. We shall suppose that the μ_i are set in an increasing order. Then μ_1 is a simple eigenvalue associated to an eigenfunction, w_1, which is strictly positive in $]0,1[$. Then we have the:

Theorem 3.5 (Principle of linearized stability) - With the above assumptions on r, u_o, and v, we have the following:

(i) (Existence and uniqueness) (3.41) has an unique solution on $[0,+\infty[$ with initial value u_o. It is a continuous function $u:[0,+\infty[\to L^2(0,1)$ such that $u(0) = u_o$ and on $]0,+\infty[$ we have $u(t) \in H^2(0,1) \cap H_o^1(0,1)$.

(ii) (Stability) If the spectrum of L lies in $]\beta,+\infty[$ for some $\beta > 0$, then v is asymptotically stable. More precisely, there exist $\delta > 0$ and $M \geq 1$ such that if $\|u_o - v\| < \delta$ then on $0 \leq t < \infty$ the solution of (3.41) satisfies:

$$\|u(t) - v\| \leq 2Me^{-\beta t} \|u_o - v\| .$$

(iii) (Asymptotic behavior) Assume that $0 < \mu_1 < \gamma < \mu_2$. Then there exists $\delta > 0$ so that $\|u_o - v\| \leq \delta$ implies:

$$u(t) = v + K(u_o)e^{-\mu_1 t} + \varepsilon(t)$$

where:

$$\|\varepsilon(t)\| \leq C \|u_o - v\| e^{-\gamma t}.$$

Here $K(v) = 0$ and $K(\cdot)$ is a continuous map from a neighborhood of v in $H_o^1(0,1)$ into the one-dimensional space $N(L - \mu_1 I) = \text{span}\,\{w_1\}$, and:

$$K(u_o) = E_1(u_o - v) + O(\|u_o - v\|^2) \text{ as } u_o \to v,$$

where E_1 is the projection of $H_o^1(0,1)$ onto span $\{w_1\}$:

$$E_1(\phi) = ((\phi, w_1))w_1 \text{ if } \|w_1\| = 1.$$

(iv) <u>(Instability)</u> Assume $\mu_1 < 0$. Then the equilibrium solution v is unstable.

For the above statement we rely heavily on the work of D. Henry [16], to which we refer for more details, and in particular for the proof of similar results in a much more general framework.

Coming back to the case of our substrate inhibited system, where $f(u) = -u'' + \sigma F(u + s_o)$, one has a linearized operator $f_u(v): w \to -w'' + \sigma F'(v + s_o)w$ and its eigenvalues correspond in the above way to the stability of the equilibrium solution v, provided the assumptions of Theorem 3.5 hold.

The only property which may not be evident is (3.44). But:

$$E(x) = r(v(x) + z(x)) - r(v(x)) - r'(v(x))z(x)$$

$$= \frac{1}{2} r''(v(x) + \theta(x)z(x))z^2(x) \qquad (0 < \theta(x) < 1)$$

and, if $|r''(\xi)| \leq M$ for every $\xi \in R$, we have:

$$|E(x)| \leq 1/2\, Mz^2(x) \leq 1/2\, M|z|_o^2 \leq C\,\|z\|^2$$

where: $|z|_o = \max\limits_{0 \leq x \leq 1} |z(x)|$, C is some constant and the last inequality is a consequence of the continuous imbedding of $H_o^1(0,1)$ into $C([0,1])$.

This completes the verification of Theorem 3.5 assumptions. However we obtained precise results in Section 3.4 about the least eigenvalue of the linearized operator $f_u(v)$.

Namely it is strictly positive on the upper and lower arcs of the S-shaped curve, those which correspond to maximal and minimal solutions, and it changes sign at turning points. Although a priori everything can happen between the two turning points, it results from the numerical exploration of the whole curve of solutions that between these turning points the least eigenvalue remains negative and no other eigenvalue crosses 0. Thus the situation is quite clear:

the steady states on the upper and lower arcs are stable, whereas those on the intermediate curve are unstable. The only steady states having a physical existence are those which are extremal, either maximal or minimal: only these can be experimentally observed, in particular on the hysteresis loop.

3.6 Justification of the Quasi-steady-state Hypothesis

In Section 3.1 we modeled the 0-dimensional system by the evolution equation (3.9), where s_o was a time varying parameter, first increasing, then decreasing. Then we dropped the time derivative, thus obtaining equation (3.11). The so-called quasi-steady-state hypothesis assumes the solution $s(\tau)$ of the differential equation (3.9) to nearly satisfy the algebraic equation (3.11). The aim of this section is to make this statement more precise.

The underlying assumption is that the diffusion-reaction process relaxes faster than s_o is altered, namely that θ is small with respect to T, T being the period of time during which s_o increases from 0 to its maximum value. Thus it is appropriate to keep the original time scale t instead of τ in (3.9), which gives:

$$(3.45)\quad \begin{cases} \varepsilon y'(t) = \lambda(t) - y(t) - \rho F(y(t)), & 0 < t \leq T, \\ y(0) = 0 \end{cases}$$

where $\varepsilon = \theta = 1/P$, $t = \varepsilon\tau$, $y(t) = s(\tau)$ and $\lambda(t) = s_o(\tau)$.

Replacing ε by 0 in (3.45) gives:

$$(3.46)\quad \lambda(t) - y(t) - \rho F(y(t)) = 0, \qquad 0 \leq t \leq T.$$

We have seen in Section 3.1 that equation (3.46) may define multiple solutions y(t). So we define z(t) as the minimal solution of:

$$(3.47)\quad \lambda(t) - z(t) - \rho F(z(t)) = 0, \qquad 0 \leq t \leq T.$$

Our aim is to justify the replacement of equation (3.45) by equation (3.46). This will be achieved by Proposition 3.11. Since the solution y of (3.45) depends on ε, which takes various values in the following, we shall denote it sometimes $y(t,\varepsilon)$ instead of y(t). When

ε takes sequential values ε_n, then y_n will denote the function $t \to y(t,\varepsilon_n)$.

It is useful to consider the two points $P = P(t,\varepsilon) = (\lambda(t) - \varepsilon y'(t,\varepsilon), y(t,\varepsilon))$ and $Q = Q(t) = (\lambda(t), z(t))$. Here $y'(t,\varepsilon)$ denotes the derivative of y with respect to t. Both points P and Q move on the curve OABCD (Figure 3.12) since their coordinates satisfy the equation of this curve, $x = y + \rho F(y)$. The path followed by the point Q as t varies from 0 to T is OACD, the last point D being attained at time T. τ will denote the "jump time", at which the point Q jumps from A to C. Thus $\lambda(\tau)$ is the abscissa of the turning point A. Let η be a strictly positive number, arbitrarily small. We have the following proposition:

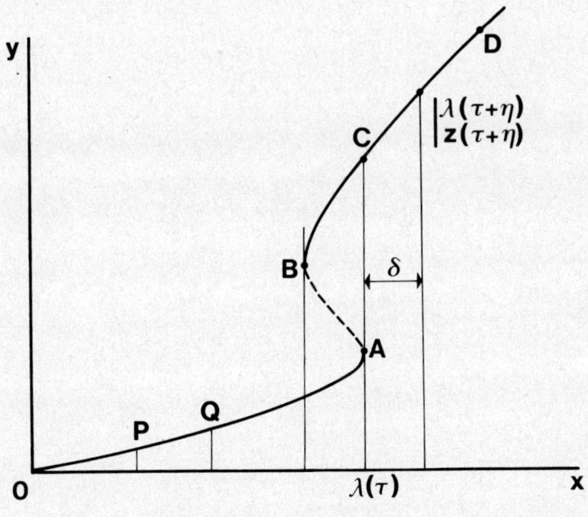

Fig. 3.12

Proposition 3.11 - Let the ε-system be defined by:

$$(3.48) \quad \begin{cases} \varepsilon y'(t) + y(t) + \rho F(y(t)) = \lambda(t), & 0 < t \leq T, \\ y(0) = 0 \end{cases}$$

and the 0-system by:

$$(3.49) \quad \begin{cases} z(t) \text{ is the minimal solution of} \\ z(t) + \rho F(z(t)) = \lambda(t), & 0 \leq t \leq T, \end{cases}$$

where,

(3.50) $\lambda \in H^1(0,T), \quad \lambda'(t) \geq 0, \quad \lambda(0) = 0.$

Then when $\varepsilon \to 0$,

(3.51) $y(t) \to z(t)$ for every $t \neq \tau$,

(3.52) $y \to z$ in $H^1(0,\tau-\eta)$ weak,

(3.53) $y \to z$ in $H^1(\tau+\eta,T)$ weak,

(3.54) $y \to z$ in $L^2(0,T)$ strong.

The proof of Proposition 3.11 will result from a series of Lemmas.

Lemma 3.2 - If, at time t_o, $y(t_o) \leq z(t_o)$, and if, during the time interval (t_o,t_1), both points P and Q remain on arc OA, then $y(t) < z(t)$, that is P is on the left of Q, for $t_o < t < t_1$. The same property holds if, instead of being on arc OA, both points P and Q belong to arc BCD.

Proof: Let $w = z - y$. w satisfies:

$$(3.55) \quad \begin{cases} \varepsilon w' + \left(\displaystyle\int_o^1 (1 + \rho F'(y + \theta w))d\theta \right) w = \varepsilon z' \\ w(t_o) \geq 0. \end{cases}$$

Since, when both points P and Q are on arc OA, (or BCD),

$$1 + \rho F'(y + \theta w) > 0 \text{ and } z' > 0, \text{ then } w > 0.$$

Lemma 3.3 - When $t \in]0,\tau[$, P is between O and Q.

Proof: We apply Lemma 3.2 with $t_o = 0$. As $y(0) = z(0) = 0$, the only property to check is that both points P and Q remain on arc OA during the time interval $(0,\tau)$. It is the case for Q by definition. P is on arc OA at least at the beginning, for t small enough, since the coordinates of P are continuous functions of t. During this period, from Lemma 3.2, $y(t) < z(t)$ and P remains between O and Q.

On the other hand suppose P crosses Q on arc OA at some time $t_1 < \tau$. Since the function $t \to y(t) - z(t)$ vanishes at $t = t_1$ by passing from negative to positive values, $y'(t_1) - z'(t_1)$ is non-negative, and $\varepsilon y'(t_1) = 0$. Hence $0 < z'(t_1) \leq y'(t_1) = 0$, which is contradictory.

Just after the jump of Q from A to C, P is still somewhere between O and A. Then, since its coordinates are continuous functions of t, P moves continuously on the curve OABCD. As long as P remains on arc OABC, $y(t) + \rho F(y(t)) \leq \lambda(\tau)$, where $\lambda(\tau)$ is the abscissa of A. Thus if $P \in$ arc OABC and $Q \in$ arc CD, we have:

$$y'(t) = \lambda(t) - (y(t) + \rho F(y(t))) \geq \lambda(t) - \lambda(\tau) > 0,$$

and P moves upwards.

<u>Lemma 3.4</u> - Let $t > \tau$. For ε small enough, $P(t,\varepsilon) \in$ arc CD.

<u>Proof</u>: If not, there exists a sequence $\varepsilon_n \to 0$ such that $P(t,\varepsilon_n) \in$ arc OABC. Let $\eta = (t - \tau)/2$ and $\delta = \lambda(\tau + \eta) - \lambda(\tau)$ (Figure 3.12). It results from the above that, for $\tau + \eta < \theta < t$, $P(\theta,\varepsilon_n) \in$ arc OABC with $y(\theta,\varepsilon_n) < y(t,\varepsilon_n)$, so that:

$$\lambda(\theta) - \varepsilon_n y'(\theta,\varepsilon_n) \leq \lambda(\tau) < \lambda(\tau + \eta) < \lambda(\theta)$$

and

$$y'(\theta,\varepsilon_n) > \delta/\varepsilon_n.$$

It follows that:

$$y(t,\varepsilon_n) - y(\tau+\eta,\varepsilon_n) = \int_{\tau+\eta}^{t} y'(\theta,\varepsilon_n)d\theta > \eta\delta/\varepsilon_n,$$

which is arbitrarily large. However this contradicts the fact that $y(t,\varepsilon_n) - y(\tau+\eta,\varepsilon_n)$ is less than the ordinate of C. This completes the proof of the Lemma.

Let η be a positive number, arbitrarily small. We wish to compare the functions $y(t,\varepsilon)$ and $z(t)$ on intervals $(0,\tau-\eta)$ and $(\tau+\eta,T)$, when ε is sufficiently small for $P(\tau+\eta,\varepsilon)$ to belong to arc CD. As Q is on the right of P when P crosses C, Q remains on the right of P during the time interval (τ,T). The reasoning is the same as in the proof of Lemma 3.3, by employing Lemma 3.2.

Lemma 3.5 - The family of functions $y(\cdot,\varepsilon)$ when ε is small enough remains in a bounded set of $H^1(0,\tau-\eta)$. (resp. $H^1(\tau+\eta,T)$).

Proof: Let (t_o,t_1) be either interval $(0,\tau-\eta)$ or $(\tau+\eta,T)$. We have, upon multiplying (3.55) by w, integrating and using the fact that, on (t_o,t_1), $1 + \rho F'(y) \geq \alpha > 0$,

$$\frac{1}{2}\,\varepsilon w^2(t) + \int_{t_o}^{t} w^2 dt \leq \varepsilon \int_{t_o}^{t} z'w\,dt \leq \frac{\varepsilon\alpha}{2} \int_{t_o}^{t} w^2 dt +$$

$$\frac{\varepsilon}{2\alpha} \int_{t_o}^{t} z'^2 dt \quad \text{where } t_o < t \leq t_1$$

and the last inequality results from the well known formula:

$$ab \leq \frac{\alpha}{2}\,a^2 + \frac{1}{2\alpha}\,b^2$$

whatever a and b. Thus, if $\varepsilon < 1$,

$$\varepsilon w^2(t) + \alpha \int_{t_o}^{t} w^2 dt \leq \frac{1}{\alpha} \int_{t_o}^{t_1} z'^2 dt, \qquad t_o < t \leq t_1.$$

It follows that:

(3.56) $\varepsilon^{1/2} w \in$ bounded set of $L^\infty(t_o,t_1)$

(3.57) $w \in$ bounded set of $L^2(t_o,t_1)$.

Proof of Proposition 3.11 - Let (t_o,t_1) be either $(0,\tau-\eta)$ or $(\tau+\eta,T)$. We can consider a sequence $\varepsilon_n \to 0$ and such that:

$y(\cdot,\varepsilon_n) \to u$ in $H^1(t_o,t_1)$ weak,

$y(\cdot,\varepsilon_n) \to u$ in $L^2(t_o,t_1)$ strong and a.e.

$\varepsilon_n^{1/2}\, y'(\cdot,\varepsilon_n) \to \xi$ in $L^\infty(t_o,t_1)$ weak* which means that:

$$\int_{t_o}^{t_1} \varepsilon_n^{1/2} \, y'(t,\varepsilon_n)\phi(t)dt \to \int_{t_o}^{t} \xi(t)\phi(t)dt \quad \text{for every}$$

$\phi \in L^1(t_o,t_1)$.

But F being Lipschitz-continuous,

$$|F(y_n(t)) - F(u(t))| \leq L|y_n(t) - u(t)|$$

so that:

$$\int_{t_o}^{t_1} |F(y_n(t)) - F(u(t))|^2 dt \leq L^2 \int_{t_o}^{t_1} |y_n(t) - u(t)|^2 dt \to 0.$$

and

$$F(y_n) \to F(u) \quad \text{in} \quad L^2(t_o,t_1).$$

For any function $v \in L^2(t_o,t_1)$ we have:

$$\varepsilon_n^{1/2} \int_{t_o}^{t_1} \varepsilon_n^{1/2} \, y_n' v \, dt + \int_{t_o}^{t_1} y_n v \, dt + \rho \int_{t_o}^{t_1} F(y_n) v \, dt = \int_{t_o}^{t_1} \lambda v \, dt.$$

As $n \to \infty$,

$$\int_{t_o}^{t_1} \varepsilon^{1/2} \, y_n' v \, dt \to \int_{t_o}^{t_1} \xi v \, dt,$$

$$\int_{t_o}^{t_1} y_n v \, dt \to \int_{t_o}^{t_1} uv \, dt,$$

and $$\int_{t_o}^{t_1} F(y_n) v \, dt \qquad \int_{t_o}^{t_1} F(u) v \, dt,$$

so that:

$$\int_{t_o}^{t_1} (u + \rho F(u) - \lambda)v \, dt = 0 \qquad \text{for every } v \in L^2(t_o,t_1),$$

whence:

$$(3.58) \quad u + \rho F(u) = \lambda.$$

But we know that $y(t,\varepsilon_n) \leq z(t)$.

Thus:

$$u(t) = \lim_{\varepsilon_n \to 0} y(t,\varepsilon_n) \leq z(t),$$

and, as z is the minimal solution of (3.58), $u(t) = z(t)$, $t \in (t_o,t_1)$.

(3.51) results from the fact that η is arbitrarily small. (3.54) results from (3.51), the fact that $y(t,\varepsilon_n) \leq z(t)$, and the Lebesgue Theorem.

Remark 3.2 - The biochemical interpretation of the above numerical and mathematical results is that, even in very simple biochemical systems where diffusion and enzyme reaction interact, multiple stable steady-states may exist, together with an hysteresis phenomenon as some environmental parameter varies. It has been suggested that such hysteresis effects in biochemical systems are adequate to account for short-term memory [17],18,19].

References

[1] Kubicek, M., Dependence of solution of nonlinear systems on a parameter, ACM Transactions on Mathematical Software, Vol. 2,#1 (March 1976) 98-107.

[2] Crandall, M.G. and Rabinowitz, P.H., Bifurcation, perturbation of simple eigenvalues, and linearized stability, Arch. Rat. Mech. Anal., 52 (1973) 161-180.

[3] Hardt, S., Naparstek, A., Segel, L.A., and Caplan, S.R., Spatio-temporal structure formation and signal propagation in a homogeneous enzymatic membrane, in: Thomas, D. and Kernevez, J.P. (eds.), Analysis and Control of Immobilized Enzyme Systems (North-Holland, Amsterdam, 1976).

[4] Naparstek, A., Romette, J.L., Kernevez, J.P., and Thomas, D., Memory in enzyme membranes, Nature, 249 (1974) 490-491.

[5] Crandall, M.G. and Rabinowitz, P.H., Bifurcation from simple eigenvalues, Journal of Functional Analysis, 8 (1971) 321-340.

[6] Crandall, M.G. and Rabinowitz, P.H., Some continuation and variational methods for positive solutions of nonlinear elliptic eigenvalue problems, Arch. Rat. Mech. Anal., 58 (1975) 207-218.

[7] Brauner, C.M. and Nicolaenko, B., Singular perturbation, multiple solutions, and hysteresis in a nonlinear problem, in: Singular Perturbation Bound. Layer Theory, Proc. Cong. Lyon 1976, Lecture Notes in Mathematics 594 (Springer, Berlin, 1977).

[8] Mignot, F. and Puel, J.P., Sur une classe de problèmes non-linéaires avec non linéarité positive, croissante, convexe, Rapport N° 75018, Analyse Numérique, Université Paris 6 (1978).

[9] Dieudonné, J., Foundations of Modern Analysis (Academic Press, New Jersey, 1960).

[10] Crandall, M.G., An introduction to constructive aspects of bifurcation and the implicit function theorem, in: Rabinowitz (ed.) Applications of Bifurcation Theory (Academic Press, New York, 1977).

[11] Krein, M.G. and Rutman, M.A., Linear operators leaving invariant a cone in a Banach space, Trans. A.M.S., Series 1, 10 (1962) 199-325.

[12] Rabinowitz, P.H., Some aspects of nonlinear eigenvalue problems, Rocky Mountain Journal of Mathematics, Vol. 3, N°2 (1973) 161-202.

[13] Leray, J. and Schauder, J., Topologie et équations fonctionnelles, Ann. Sci. Ecole Norm. Sup. 3, Vol. 51 (1934) 45-78.

[14] Sattinger, D.H., Topics in stability and bifurcation theory, Lecture Notes in Mathematics, N°309 (Springer Verlag, Berlin, 1973).

[15] Brauner, C.M. and Nicolaenko, B., Perturbation singulière, solutions multiples et hystérésis dans un problème de biochimie, Cras Série 1 (1976).

[16] Henry, D., Geometric theory of semilinear parabolic equations, Lecture Notes, University of Kentucky (1976).

[17] Changeux, J.P. and Thiery, J., in: Järnefelt J.(ed.), Regulatory Function of Biological Membranes (Elsevier, New York, 1968).

[18] Katchalsky, A. and Oplatka, A., Neurosciences Res. Symp., 1 (1966) 352-371.

[19] Katchalsky, A. and Spangler, R., Quarterly Rev. Biophys., 1 (1968) 127-175.

CHAPTER 4

ASSEMBLAGES OF CELLS AND BIFURCATIONS

In our survey of the various phenomena induced by interacting diffusion and reaction in enzyme systems, we first observed the existence of concentration gradients in the profiles within the membranes. This phenomenon implies the necessity, when discussing membrane behavior, to take into account not only the external concentrations, but also the internal values which may be much smaller. Then we have seen how a "scalar" phenomenon, enzyme reaction, can cause a "vectorial" effect, glucose transport, in a glucose pump where transport against the concentration gradient is effected from an external reservoir at low concentration to an internal compartment at high concentration. Lastly, we studied a system endowed with an hysteretic behavior due to the existence of multiple steady states and able to act as a memory and store information. The goal of this chapter is, on one hand, to study a type of system interesting in itself, that is assemblages of cells, and on the other hand, to explain by a simple example a type of phenomenon that we shall encounter frequently in succeeding chapters, bifurcation. In particular, we shall study the solutions (s_1,s_2) of the system:

$$(4.1)\quad \begin{cases} 2s_1 - s_2 + \rho F(s_1) = s_o \\ -s_1 + 2s_2 + \rho F(s_2) = s_o \end{cases}$$

as the parameter s_o varies. Here F(s) is, as in Chapter 3, the substrate inhibited rate expression:

$$(4.2)\quad F(s) = s/(1 + |s| + ks^2).$$

Equations (4.1) represent, in the case N = 2, the steady states of the system:

$$
(4.3)\quad \begin{cases}
\dfrac{ds_1}{dt} + 2s_1 - s_2 + \rho F(s_1) = s_o \\
\cdots\cdot \\
\dfrac{ds_i}{dt} - s_{i-1} + 2s_i - s_{i+1} + \rho F(s_i) = 0, \qquad 2 \le i \le N-1, \\
\cdots\cdot \\
\dfrac{ds_N}{dt} - s_{N-1} + 2s_N + \rho F(s_N) = s_o
\end{cases}
$$

Here $s_1, s_2, \ldots, s_N$ are substrate concentrations in a linear array of cells containing a substrate inhibited enzyme, each cell being in contact with two neighbors, except for the two cells at the ends of the array which are in contact with well-stirred reservoirs at specified concentration s_o (Figure 4.1). Such assemblages of cells with substrate inhibition kinetics have been studied by Bunow and Colton [1], who so describe the biological interest of their model: "In the tissues of metazoan organisms, some cells have direct access to compartments or reservoirs which supply the substrates or remove the products; for other cells the access is diminished by the presence of neighboring cells through which these substances must pass in order to reach the interior of the tissue. The intervening cells may contain the same enzymes which catalyze reactions involving these substrates, thereby reducing the substrate concentration available to interior cells. These phenomena may be expected to change the apparent dependence of enzymatic reaction rates in an assemblage of cells upon the substrate concentration in the surrounding medium, in comparison either to the observed kinetics of single cells or to that of enzymes free in solution".

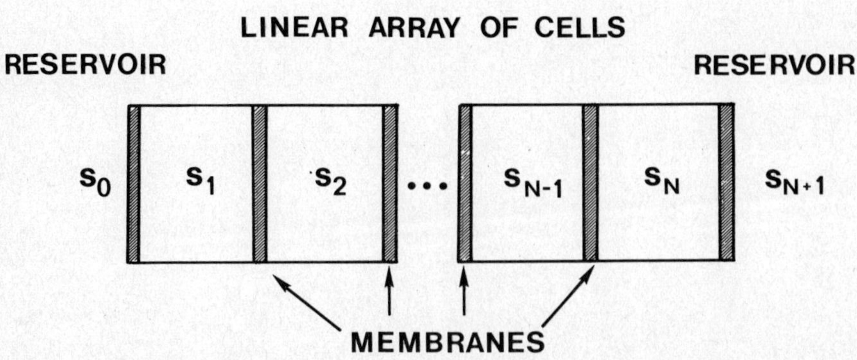

Fig. 4.1

An additional interest of Bunow and Colton's work is to show that over prescribed ranges of reservoir concentration s_o, multiple steady states can occur, some of which are characterized by asymmetric profiles of concentration and reaction rates across the array.

Figure 4.2 shows the first component s_1 of the solution (s_1,s_2) of (4.1) as a function of the external concentration s_o. In addition to a smooth curve of "symmetric" solutions, that is satisfying $s_1 = s_2$, represented by the curve OABCDZ, we shall find branches of nonsymmetric solutions $(s_1 \neq s_2)$ emerging from this curve at two bifurcation points B and C (BEFC and BGHC).

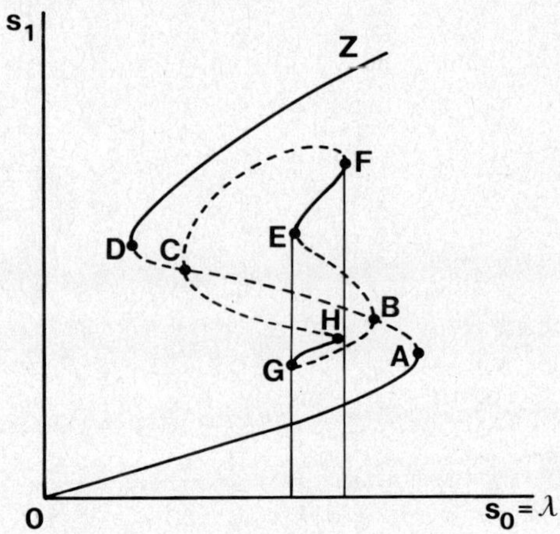

Fig. 4.2

The outline for the remainder of this chapter is as follows. Section 4.1 describes multiple cell assemblages. Because the analysis and exposition are simplest when there are only two cells, in Section 4.2 we restrict our attention to this case. There is a well developed theory which permits the qualitative description of the solutions for equations:

(4.4) $\quad f(\lambda,u) = 0$

in the vicinity of a bifurcation point (λ_o,u_o). In this chapter we restrict our analysis to mappings $f : R^+ \times R^N \to R^N$ twice continuously differentiable with respect to both λ and u in some neighborhood of the bifurcation point (λ_o,u_o). In Section 4.3 we set the framework for further discussion, specify conditions for bifurcation to occur, prove the existence of bifurcations and study exchange of stability at bifurcation points. In Section 4.4 we describe numerical methods for computing bifurcation points and bifurcation branches. In both sections we rely upon the methods of H.B. Keller and his group [2,3,4,5,6].

Finally Section 4.5 deals with the case of slightly different boundary values:

$$(4.5)\quad \begin{cases} 2s_1 - s_2 + \rho F(s_1) = s_o \\ -s_1 + 2s_2 + \rho F(s_2) = s_o + \varepsilon. \end{cases}$$

Here ε is a small "imperfection" the influence of which on the family of solutions is analyzed. More precisely, we explain why there is bifurcation breaking.

4.1 Multiple Cell Assemblages

Figure 4.1 shows an array of N identical cells, each containing an enzyme catalyzing a reaction inhibited by high substrate concentrations, with rate expression:

$$R(S) = V_M S/(K_S + S(1 + S/K_{SS})),$$

V_M, K_S, and K_{SS} being the same in each cell. Furthermore, suppose that the cells communicate with one another by diffusion across their common boundary. Across the membrane separating cells i and i-1 the rate of substrate transport is $PA(S_{i-1} - S_i)$, where P is the diffusive permeability and A is the contact surface area. Figure 4.1 shows a pair of reservoirs at the ends of the array, at specified concentrations S_o and S_{N+1}, which provide a continuous source of substrate. Mass balance for substrate in the ith cell is given by:

$$(4.6)\quad V\frac{dS_i}{dt} = PA(S_{i-1} - S_i) + PA(S_{i+1} - S_i) - VR(S_i)$$

where V is the cell volume. By introducing a reduced substrate concentration $s = S/K_S$ and a reduced time $\tau = t\,PA/V$, equation (4.6) may be rewritten in dimensionless form:

(4.7) $$\frac{ds_i}{d\tau} = s_{i-1} - 2s_i + s_{i+1} - \rho F(s_i)$$

where F(s) is given by (4.2) with $k = K_S/K_{SS}$ and:

(4.8) $$\rho = \frac{V_M}{K_S}\frac{V}{PA} .$$

The composition of the cells in the steady state is governed by:

(4.9) $$\begin{cases} 2s_1 - s_2 + \rho F(s_1) = s_o \\ -s_{i-1} + 2s_i - s_{i+1} + \rho F(s_i) = 0 & 2 \leq i \leq N-1 \\ -s_{N-1} + 2s_N + \rho F(s_N) - s_{N+1} \end{cases}$$

In Section 4.2 we shall be in the case N = 2 and $s_{N+1} = s_o$. For the numerical experiments of Section 4.4 we shall take $N \geq 2$ and $s_{N+1} = s_o$. In Section 4.5 we shall have N = 2 but $s_{N+1} = s_o + \varepsilon$.

4.2 Two Cell Assemblage

Consider the steady state equations:

(4.10) $$\begin{cases} 2s_1 - s_2 + \rho F(s_1) = s_o \\ -s_1 + 2s_2 + \rho F(s_2) = s_o \end{cases}$$

as s_o varies. A first branch of solutions consists of "symmetric" profiles $s_1 = s_2$ obtained by replacing s_1 and s_2 by s into (4.10), which gives:

(4.11) $$s + \rho F(s) = s_o .$$

We already encountered equation (4.11) in Section 3.1, and saw that, in the (s_o,s) plane, its solutions are located on an S-shaped curve. Thus we obtain a family of solutions of (4.10) parametrized by s,

$$s_o = s + \rho F(s), \quad s_1 = s, \quad s_2 = s.$$

As s varies from 0 to $+\infty$, the s_1 component is represented against s_o by the S-shaped curve OABCDZ of Figure 4.2. The stability of these "symmetric" steady-states depends on the eigenvalues of the linearized operator:

$$\begin{bmatrix} 2 + \rho F'(s_1) & -1 \\ -1 & 2 + \rho F'(s_2) \end{bmatrix}$$

for $s_1 = s_2 = s$. These eigenvalues are $\mu_1 = 1 + \rho F'(s)$ and $\mu_2 = 3 + \rho F'(s)$, and we suppose ρ sufficiently large for μ_1 and μ_2 to be able to vanish for some values of s. More precisely $\mu_1 = 0$ when $F'(s) = -1/\rho$, which occurs for $s = t_1$ and $s = t_2$ (Figure 4.3), thus giving the turning points A and D, and $\mu_2 = 0$ when $F'(s) = -3/\rho$, which occurs for $s = b_1$ and $s = b_2$, corresponding to the bifurcation points B and C.

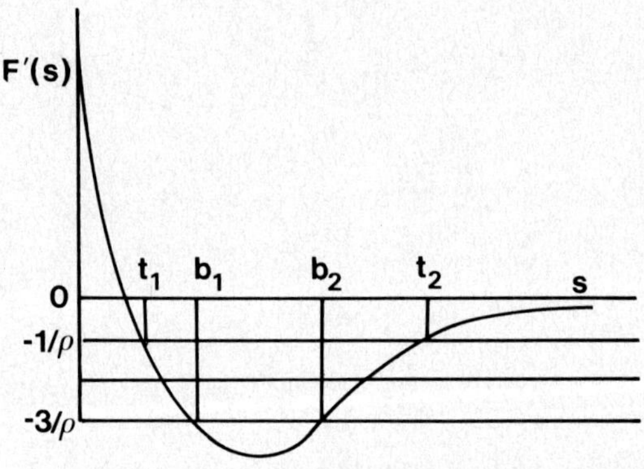

Fig. 4.3

The stability analysis of the "symmetric" states is straightforward. They are stable when the leading eigenvalue μ_1 is positive, that is when $0 \leq s < t_1$ or $s > t_2$, and unstable when $t_1 < s < t_2$. Thus the middle states, which occur between A and D, in the region of negative slope, are unstable whereas the lower and upper states, on limbs OA and DZ of the sigmoid curve, are stable.

It is a fairly easy exercise to check that points A and D are turning points, by verifying the hypotheses of Theorem 3.2. But we are mainly interested by the status of points B and C, more precisely to know why branches of solutions emanate from these points and how they can be calculated. Before discussing mathematical and numerical aspects of bifurcations in Sections 4.3 and 4.4, let us change notations and specify the properties which will be used.

Notations - It is more appealing to denote the variables by names reminiscent of their origin, for example s_o, s_1, and s_2 for substrate concentrations. However, in order to cast problem (4.10) into a more general framework, we perform a dramatic change of variables:

$$\lambda = s_o,\ u_1 = s_1,\ u_2 = s_2.$$

Let $u = (u_1,u_2)$ and $f : R^+ \times R^2 \to R^2$ be the mapping:

$$(4.12)\quad (\lambda,u) \to (2u_1 - u_2 + \rho F(u_1) - \lambda,\ -u_1 + 2u_2 + \rho F(u_2) - \lambda).$$

Furthermore s_o will designate in Sections 4.3 and 4.4, one of the two solutions b_1 and b_2 of $3 + \rho F'(s_o) = 0$ (Figure 4.3). (λ_o,u_o) will be the corresponding bifurcation point (B or C) defined by $\lambda_o = s_o + \rho F(s_o)$ and $u_{o1} = u_{o2} = s_o$.

$f_u^0 = f_u(\lambda_o,u_o)$ and $f_\lambda^0 = f_\lambda(\lambda_o,u_o)$ will be respectively the Jacobian matrix of f and the derivative of f with respect to λ at (λ_o,u_o):

$$f_u^0 = \begin{bmatrix} -1 & -1 \\ -1 & -1 \end{bmatrix} \qquad \text{and } f_\lambda^0 = \begin{bmatrix} -1 \\ -1 \end{bmatrix}.$$

ϕ will be an eigenvector of f_u^0 associated to the zero eigenvalue:

$$\phi = \begin{bmatrix} 1 \\ -1 \end{bmatrix}.$$

ψ^* will be an eigenvector of f_u^{0*}, the transpose of f_u^0. In fact, since here $f_u^{0*} = f_u^0$, we shall have $\psi^* = \phi^*$ (transpose of ϕ) = $[1\ \ -1]$.

Properties - The null space $N(f_u^0)$ and range $R(f_u^0)$ of f_u^0 satisfy:

$$\dim N(f_u^0) = \operatorname{codim} R(f_u^0) = 1,$$

$$N(f_u^0) = \text{span } \{\phi\},\ N(f_u^{0*}) = \text{span } \{\psi^*\}$$

$$v \in R(f_u^0) \Leftrightarrow \psi^* v = 0$$

where ψ^* and v are understood to be respectively row and column vectors, and $\psi^* v$ the scalar product of these vectors.

The vector f_λ^0 is such that:

$$\psi^* f_\lambda^0 = 0.$$

This is precisely the property which will permit us to recognize the turning points and bifurcation points as being different.

At either type of point f_u^0 is singular, but at turning points $\psi^* f_\lambda^0 \neq 0$ (Theorem 3.2) whereas at bifurcation points $\psi^* f_\lambda^0 = 0$. Compare for example the calculated values of $\psi^* f_\lambda^0$ at turning points and at bifurcation points in Figure 4.4.

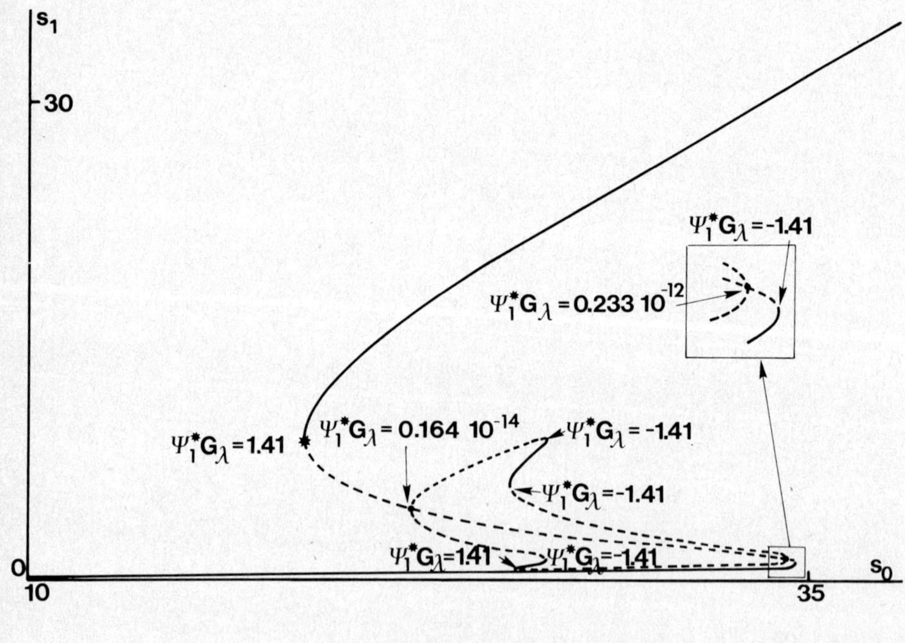

Fig. 4.4

4.3 Nonisolated Solutions and Bifurcation

Consider the equation:

(4.13) $f(\lambda,u) = 0$

where $f : R^+ \times R^N \to R^N$ is a twice continuously differentiable mapping. We deliberately restrict ourselves to R^N in this Chapter and the next because the states u of the systems therein considered lie in finite dimensional spaces, and also because R^N is the usual framework for numerical approximations of infinite dimensional problems such as encountered later. However, since most of the concepts and results of this section are extendable to mappings $f : R^+ \times B \to B$ where B is a real Banach space, R^N will be denoted B. Keener [7] gives the following definitions of branching and bifurcation points which, although somewhat restrictive, encompass all of the phenomena we discuss in this book.

Definition 4.1 - Suppose (λ_o,u_o) is a solution of (4.13). Then (λ_o,u_o) is a branching point of (4.13) if in every open neighborhood of (λ_o,u_o) there exist distinct solutions (λ_1,u_1) and (λ_2,u_2) of (4.13) with $\lambda_1 = \lambda_2$.

It is the case of points A,B,C,D,E,F,G,H in Figure 4.2. However there is a difference of nature between turning points A,D,E,F,G and H and bifurcation points B and C. For a branching point to be a bifurcation point, we need an additional property.

Definition 4.2 - A branching point is a bifurcation point if, furthermore, every open neighborhood of (λ_o,u_o) contains at least one solution of (4.13) for $\lambda > \lambda_o$ and also for $\lambda < \lambda_o$.

Definition 4.3 - By a smooth branch or arc Γ of solutions for (4.13) we mean a twice continuously differentiable mapping $\Gamma : s \to (\lambda(s), u(s))$ of a compact interval $[\alpha,\beta] \subset R$ into $R^+ \times B$, satisfying $f(\lambda(s), u(s)) = 0$. In other words it is a smooth one parameter family of solutions for (4.13), $(\lambda(s),u(s))$, s describing the interval $[\alpha,\beta]$.

Let:

$$f_u(\lambda,u) = \left(\frac{\partial f_i}{\partial u_j}(\lambda,u)\right)_{1 \le i,j \le N}$$

be the Jacobian matrix of f, and $f_u^0 = f_u(\lambda_o,u_o)$.

If f_u^0 is nonsingular, the implicit function theorem implies that there are neighborhoods U of λ_o in R^+ and V of u_o in B such that there is exactly one function $u : U \to V$ satisfying $f(\lambda,u(\lambda)) = 0$ for $\lambda \in U$. Moreover, the mapping f being twice continuously differentiable on $R^+ \times B$, u is twice continuously differentiable and the solution (λ_o,u_o) can be continued into exactly one smooth branch of solutions for (4.13).

<u>Definition 4.4</u> - A solution pair (λ_o,u_o) at which f_u^0 is non singular (resp. singular) is called an isolated (resp. a nonisolated) solution of (4.13).

Of course a necessary condition for a solution (λ_o,u_o) of (4.13) to be a branching solution is that it be a nonisolated solution.

Suppose that (λ_o,u_o) is a nonisolated solution of (4.13), that f_u^0 has null space $N(f_u^0)$ spanned by ϕ, and that $N(f_u^{0*})$, the null space of the transpose f_u^{0*} of f_u^0, is spanned by ψ^*. Let:

$$f_\lambda^0 = f_\lambda(\lambda_o,u_o) = \left(\frac{\partial f_i}{\partial \lambda}(\lambda_o,u_o)\right)_{1\le i\le N}.$$

The existence of a nonisolated solution is not sufficient to guarantee branching. However there are two important possibilities.

<u>Possibility 1</u> - If:

$$(4.14) \quad \psi^* f_\lambda^0 \neq 0$$

theorem 3.2 tells us that the solutions of $f(\lambda,u) = 0$ near (λ_o,u_o) form a curve $(\lambda(s),u(s)) = (\lambda_o + \tau(s), u_o + s\phi + x(s))$, where $\tau(0) = \tau'(0) = 0$, $x(0) = x'(0) = 0$ and $\psi^* x(s) = 0$. Moreover, since f is twice continuously differentiable, so is λ and the calculation of $\lambda''(0)$, resulting in formula (3.36), shows that:

$$(4.15) \quad \lambda''(0) = -\psi^* f_{uu}^0 \phi^2 / \psi^* f_\lambda^0.$$

Here $f_{uu}^0\phi^2$ is a shortened notation for $f_{uu}^0\phi\phi$, where $f_{uu}^0 vw$ is the vector of components:

$$\sum_{1\le j,k\le N} \frac{\partial^2 f_i}{\partial u_j\, u_k}(\lambda_o,u_o) v_j w_k, \quad 1 \le i \le N.$$

Thus if we suppose that (4.14) holds, (λ_o,u_o) is a <u>turning point</u>

(also referred to as a limit point or a bending point) provided:

(4.16) $\psi^* f^0_{uu}\phi^2 \neq 0.$

Depending upon cases ($\lambda''(0) < 0$ or $\lambda''(0) > 0$), there is no solution to the nonlinear problem $f(\lambda,u) = 0$ either for $\lambda > \lambda_o$ or $\lambda < \lambda_o$.

In the particular case $\lambda''(0) = 0$, then it is easy to see that (λ_o,u_o) is not a branching point or is a turning point according to whether the first nonvanishing derivative $\lambda^{(p)}(0)$ is of odd or even order.

Possibility 2 - Therefore in order that (λ_o,u_o) be a point of bifurcation,

(4.17) $\psi^* f^0_\lambda = 0$

must hold. We will see that under an additional condition which is in general satisfied, (4.17), together with the above assumptions on f^0_u, ensures the existence of switching branches at (λ_o,u_o). We rely upon [2], [3], and [5] for analyzing this situation.

We first specify the structure of the tangents to the branches at a bifurcation point.

Theorem 4.1 - Suppose:

(4.18) $f(\lambda_o,u_o) = 0$

(4.19) $\dim N(f^0_u) = \operatorname{codim} R(f^0_u) = 1,$

(4.20) $f^0_\lambda \in R(f^0_u),$

(4.21) $N(f^0_u) = \operatorname{span}\{\phi\}$, $N(f^{0*}_u) = \operatorname{span}\{\psi^*\}$.

Let ϕ_o be the unique element of B such that:

(4.22) $f^0_u\phi_o + f^0_\lambda = 0$, $\psi^*\phi_o = 0$.

We define a, b, c by:

(4.23) $a = \psi^* f^0_{uu}\phi\phi$, $b = \psi^*(f^0_{uu}\phi_o + f^0_{u\lambda})\phi$, $c = \psi^*(f^0_{uu}\phi_o\phi_o + 2f^0_{u\lambda}\phi_o + f^0_{\lambda\lambda})$.

Let $\Gamma : s \to (\lambda(s),u(s))$ be any smooth branch of (4.13) through the point (λ_o,u_o), say with $\lambda(s_o) = \lambda_o, u(s_o) = u_o$.

Then the tangent to Γ at (λ_o,u_o) is of the form:

$$(4.24) \quad \lambda'(s_o) = \alpha_o, \; u'(s_o) = \alpha_o\phi_o + \alpha_1\phi$$

where the scalars (α_o,α_1) satisfy the <u>Bifurcation Equation</u>:

$$(4.25) \quad a\alpha_1^2 + 2b\alpha_o\alpha_1 + c\alpha_o^2 = 0$$

<u>Proof</u>: (4.17) implies the existence of an unique element $\phi_o \in B$ satisfying (4.22). Differentiating the identity $f(\lambda(s),u(s)) = 0$ twice, it follows that:

$$f_u(\lambda(s),u(s))u'(s) + f_\lambda(\lambda(s),u(s))\lambda'(s) = 0$$

$$f_{uu}(s)u'^2(s) + 2f_{u\lambda}(s)\lambda'(s)u'(s) + f_u(s)u''(s) + f_{\lambda\lambda}(s)\lambda'^2(s)$$
$$+ f_\lambda(s)\lambda''(s) = 0$$

where: $f_u(s) = f_u(u(s),\lambda(s)), \ldots$

We get at $s = s_o$:

$$(4.26) \quad f_u^0u'(s_o) + f_\lambda^0\lambda'(s_o) = 0$$

$$(4.27) \quad f_u^0u''(s_o) = -[f_{uu}^0u'^2(s_o) + 2f_{u\lambda}^0\lambda'(s_o)u'(s_o) + f_{\lambda\lambda}^0\lambda'^2(s_o)]$$
$$- f_\lambda^0\lambda''(s_o)$$

where: $f_u^0 = f_u(s_o) = f_u(\lambda_o,u_o), \ldots$.

It results from (4.22) and (4.26) that $f_u^0(u'(s_o) - \lambda'(s_o)\phi_o) = 0$, that is $u'(s_o) - \lambda'(s_o)\phi_o \in N(f_u^0)$, and there exist two scalars α_o and α_1 for which (4.24) holds.

Since both $f_u^0u''(s_o)$ and $f_\lambda^0\lambda''(s_o)$ are in $R(f_u^0)$ the same must be true of the bracketed term of (4.27). Thus $\psi_1[\ldots] = 0$ must hold for this term and so using (4.24) it follows that the two scalars α_o and α_1 must satisfy the quadratic equation (4.25), where a, b, and c are given by (4.23) and this completes the proof.

The quadratic form (4.25) and its role in bifurcation theory was first pointed out in [2].

Note that the eigenvectors ϕ and ψ^*; the vector ϕ_o and the coefficients a, b, and c will be the same for any branch Γ passing through (λ_o, u_o), and are defined by quantities involving the single function f and its derivatives at that point (λ_o, u_o). In case there is a second smooth branch of (4.13), $\Gamma' : t \to (\mu(t), w(t))$, passing through the point (λ_o, u_o), say with $\mu(0) = \lambda_o$ and $w(0) = u_o$, then it follows from theorem 4.1 that the tangent to Γ' at (λ_o, u_o) is of the form:

$$\mu'(0) = \beta_o \quad \text{and} \quad w'(0) = \beta_o\phi_o + \beta_1\phi$$

where the scalars β_o and β_1 satisfy the same quadratic equation as α_o and α_1, namely the bifurcation equation:

$$a\beta_1^2 + 2b\beta_o\beta_1 + c\beta_o^2 = 0.$$

Now we specify the condition under which the roots of the bifurcation equation:

$$(4.28) \quad a\xi_1^2 + 2b\xi_o\xi_1 + c\xi_o^2 = 0$$

are distinct. As (4.28) is a homogeneous quadratic equation, it admits rays of solutions. Its roots are defined up to a scalar multiple, and two roots (α_o, α_1) and (β_o, β_1) are said to be distinct when:

$$(4.29) \quad \alpha_1\beta_o - \alpha_o\beta_1 \neq 0.$$

Lemma 4.1 - The roots of the bifurcation equation (4.28) are distinct provided:

$$(4.30) \quad \psi^*(f^0_{uu}u'(s_o) + f^0_{u\lambda}\lambda'(s_o))\phi \neq 0 \quad \text{when} \quad \lambda'(s_o) \neq 0$$

and

$$(4.31) \quad \psi^*(f^0_{uu}\phi_o + f^0_{u\lambda})\phi \neq 0 \qquad \text{when } \lambda'(s_o) = 0.$$

Proof: We already know that the quadratic (4.28) admits a nontrivial root (α_o,α_1). The other root is distinct provided:

$$a\alpha_1 + b\alpha_o \neq 0 \qquad \text{when } \alpha_o \neq 0$$

and

$$b \neq 0 \qquad \text{when } \alpha_o = 0.$$

The proof of the Lemma is completed by verifying that the left-hand sides in (4.30) and (4.31) are precisely $a\,\alpha_1 + b\,\alpha_o$ and b.

Under the assumptions of Theorem 4.1, together with (4.30) and (4.31), our goal is to prove the existence of a second curve of solutions emanating from (λ_o,u_o) and distinct from Γ. Let $\Gamma' : t \to (\mu(t),w(t))$ be such a curve, say with $\mu(0) = \lambda_o$ and $w(0) = \mu_o$. Then Theorem 4.1 tells us that we necessarily have:

$$\mu'(0) = \beta_o \quad \text{and} \quad w'(0) = \beta_o\phi_o + \beta_1\phi$$

where (β_o,β_1) is the second root of the bifurcation equation (4.28). We just noticed in Lemma 4.1 that (4.30) and (4.31) are precisely conditions ensuring that the pair of roots (α_o,α_1) and (β_o,β_1) are distinct. Thus we must have:

$$(4.32) \quad \begin{cases} \mu(t) = \lambda_o + t\xi_o(t) \\ w(t) = u_o + t\,(\xi_o(t)\phi_o + \xi_1(t)\phi) + t^2 v(t) \end{cases}$$

where:

$$(4.33) \quad \xi_o(0) = \beta_o,\ \xi_1(0) = \beta_1,\ \psi^* v(t) = 0$$

together with some normalization, say:

$$(4.34) \quad \mu'^2(t) + |w'(t)|^2 = 1.$$

In particular, at t = 0, the scaling of β_o and β_1 is specified by:

$$\beta_o^2 + |\beta_o\phi_o + \beta_1\phi|^2 = 1$$

We demonstrate the existence of Γ' using implicit function theoretic techniques, following [5], where the analysis is performed under much more general conditions.

Application of the Implicit Function Theorem: The framework is the following:

We define the function $g : R \times R \times B \times R \to B \times R \times R$ by:

$$g(t;\xi_o,\xi_1,v) = (h(t;\xi_o,\xi_1,v), \xi_o^2 + |\xi_o\phi_o + \xi_1\phi|^2 - 1, \psi_1^* v)$$

where:

$$h(t;\xi_o,\xi_1,v) = \begin{cases} \frac{1}{t^2} f(\lambda_o + t\xi_o, u_o + t(\xi_o\phi_o + \xi_1\phi) + t^2 v) & \text{if } t \neq 0, \\ f_u^0 v + \frac{1}{2}(f_{uu}^0(\xi_o\phi_o + \xi_1\phi)^2 + 2f_{u\lambda}^0(\xi_o\phi_o + \xi_1\phi)\xi_o + f_{\lambda\lambda}^0\xi_o^2) & \text{if } t = 0. \end{cases}$$

We see immediately that $g(0;\beta_o,\beta_1,v_o) = 0$ where v_o is the unique solution of:

$$(4.35) \quad \begin{cases} f_u^0 v_o + \frac{1}{2}[f_{uu}^0(\beta_o\phi_o + \beta_1\phi)^2 + 2f_{u\lambda}^0(\beta_o\phi_o + \beta_1\phi)\beta_o + f_{\lambda\lambda}^0\beta_o^2] = 0, \\ \psi^* v_o = 0 \end{cases}$$

The claim about existence and uniqueness of a solution to (4.35) follows from:

Lemma 4.2 - The bracketed term in (4.35) is in $R(f_u^0)$.

Proof: $\psi^*[\ldots] = a\beta_1^2 + 2b\beta_o\beta_1 + c\beta_o^2 = 0$

by definition of (β_o,β_1).

We now investigate the derivative of g with respect to (ξ_o,ξ_1,v) at $(0;\beta_o,\beta_1,v_o)$, applied to $(\hat{\xi}_o,\hat{\xi}_1,\hat{v})$:

$$(4.36) \quad g_{(\xi_o,\xi_1,v)}(0;\beta_o,\beta_1,v_o)\,(\hat{\xi}_o,\hat{\xi}_1,\hat{v}) =$$

$$(f_u\hat{v} + [a'\beta_1\hat{\xi}_1 + b'(\beta_1\hat{\xi}_o + \beta_o\hat{\xi}_1) + c'\beta_o\hat{\xi}_o],$$

$$2(\beta_o\hat{\xi}_o + (\beta_o\phi_o + \beta_1\phi, \hat{\xi}_o\phi_o + \hat{\xi}_1\phi)), \psi^*\hat{v})$$

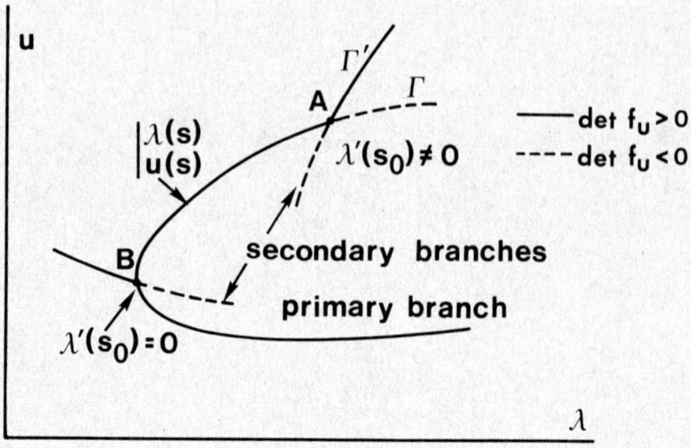

Fig. 4.5

If we assume the functions $\gamma(s)$ and $z(s)$ to be continuously differentiable in some neighborhood of s_o, we obtain by differentiating (4.38) with respect to s at $s = s_o$:

$$(4.39)\quad \begin{cases} (f^0_{uu}u'(s_o) + f^0_{u\lambda}\lambda'(s_o))w(s_o) + f^0_u w'(s_o) = \gamma'(s_o)w(s_o) + \gamma(s_o)w'(s_o) \\ w'(s_o) = z'(s_o) \\ \psi^* z'(s_o) = 0 \end{cases}$$

Acting upon the first equation in (4.39) with ψ^* and replacing $u'(s_o)$ by $\alpha_o\phi_o + \alpha_1\phi$, $\lambda'(s_o)$ by α_o, and $w(s_o)$ by ϕ we find:

$$\psi^*(f_{uu}(\alpha_o\phi_o + \alpha_1\phi) + f^0_{u\lambda}\alpha_o)\phi = \gamma'(s_o)\psi^*\phi$$

Hence:

$$(4.40)\quad \gamma'(s_o) = (a\alpha_1 + b\alpha_o)/(\psi^*\phi).$$

We distinguished two possibilities in Theorem 4.2. Either $\alpha_o \neq 0$ and $a\alpha_1 + b\alpha_o \neq 0$, or $\alpha_o = 0$ and $b \neq 0$. It follows from (4.40) that $\gamma'(s_o) \neq 0$ in the first case, whereas in the second case, since $a = 0$, $\gamma'(s_o) = 0$. Thus we have the following result:

Proposition 4.1 - Under the assumptions of Theorem 4.2, let $\gamma(s)$ be the eigenvalue of $f_u(\lambda(s),u(s))$ which vanishes at $s = s_o$. Then:

$$\gamma'(s_o) \neq 0 \qquad \text{when } \lambda'(s_o) \neq 0$$

(there is a stability change through s_o)

$$\gamma'(s_o) = 0 \qquad \text{when } \lambda'(s_o) = 0.$$

Thus although $\det(f_u(\lambda(s),u(s))$ vanishes at both points A and B in Figure 4.5, its sign changes at A but not at B (in the general case $\gamma''(s_o) \neq 0$ there is no change in stability through B).

In particular if all the eigenvalues of $f_u(\lambda(s),u(s))$ are positive before crossing A, that is if the state u(s) is stable, then after crossing the least eigenvalue is negative, and the state u(s) is unstable. There is a change of stability when $(\lambda(s),u(s))$ crosses A. On the other hand the number of negative eigenvalues is the same before and after crossing B and the stability is the same on both sides of B.

Much more can be said about the behavior of the eigenvalues which vanish at the bifurcation point as the points $(\lambda(s),u(s))$ and $(\mu(t),w(t))$ pass through this point (λ_o,u_o), namely we have the phenomenon of:

<u>Exchange of stability</u>: We describe two possibilities, which encompass all of the situations we shall be faced with in our subsequent analyses of stationary states.

<u>Possibility 1</u>: $\lambda'(s_o) \neq 0$ and $\mu'(0) \neq 0$ (Point A in Figure 4.5) Formula (4.40) applies not only to the eigenvalue $\gamma(s)$ which vanishes as $(\lambda(s),u(s))$ crosses (λ_o,u_o), but also to the eigenvalue $\delta(t)$ of $f_u(\mu(t),w(t))$ which vanishes at $t = 0$, as the point $(\mu(t),w(t))$ of Γ' crosses (λ_o,u_o):

$$\delta'(0) = (a\beta_1 + b\beta_o)/\psi^*\phi$$

Thus

$$\gamma'(s_o)\delta'(0) = (a\beta_1 + b\beta_o)(a\alpha_1 + b\alpha_o)/(\psi^*\phi)^2.$$

It is easy to check that:

$$\gamma'(s_o)\delta'(0) = -\alpha_o\beta_o(b^2 - ac)/(\psi^*\phi)$$

Since the bifurcation equation possesses distinct real roots we have $b^2 - ac > 0$. Now in order to move along the two branches Γ and Γ' in the same direction, we choose α_o $(= \lambda'(s_o))$ and β_o $(= \mu'(0))$ to be of the same sign, say positive. With this choice we see that:

$$\gamma'(s_o)\delta'(0) < 0.$$

Hence the eigenvalues $\gamma(s)$ and $\delta(t)$ cross the origin in opposite directions.

In particular, if the state $u(s)$ loses stability as the point $(\lambda(s), u(s))$ on arc Γ crosses (λ_o,u_o), then $\gamma'(s_o) < 0$ and $\delta'(0) > 0$, implying that the point $(\mu(t),w(t))$ on arc Γ' gains stability. More generally, since the Jacobian matrices $f_u(\lambda(s),u(s))$ and $f_u(\mu(t),w(t))$ coincide at (λ_o,u_o), the number of strictly negative eigenvalues of the first one is transferred to the second one when crossing (λ_o, u_o).

Possibility 2:

$$\lambda'(s_o) \neq 0 \text{ and } \mu'(0) = 0,\ \mu''(0) \neq 0$$

(points B and C in Figure 4.6). We refer to [5] and [8] for a proof of the following result:

$$\lim_{t\to 0} \frac{-t\mu'(t)\gamma'(s_o)}{\delta(t)\lambda'(s_o)} = 1,$$

which implies that the full (resp. dashed) lines in Figure 4.6 correspond to the same even (resp. odd) number of negative values for $f_u(\lambda,u)$.

Application to the two-cell assemblage of Section 4.2: We already found a "trivial" branch of "symmetric" states, OABCDZ (Figure 4.2), with the parametrization:

$$\lambda = s + \rho F(s), \quad u_1 = s \text{ and } u_2 = s, \qquad s \in R^+,$$

on which points B and C satisfy the assumptions of theorem 4.2, as is easily checked. In particular $\lambda'(s_o) = 1 + \rho F'(s_o) = -2$, and $\psi^*(f^0_{uu}u'(s_o) + f^0_{u\lambda}\lambda'(s_o))\phi = -2$, so that the bifurcation equation has distinct roots.

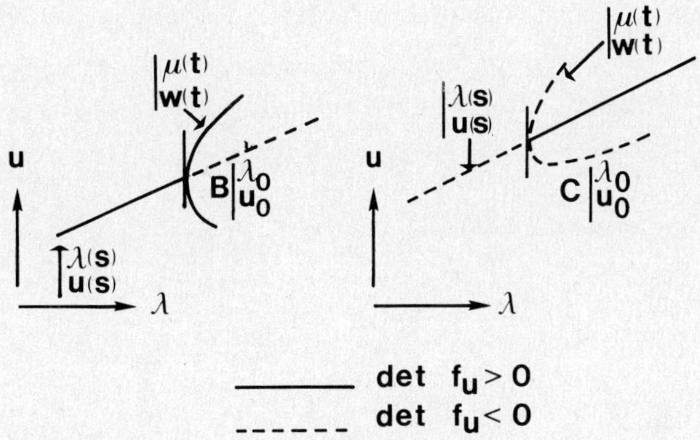

Fig. 4.6

Lastly, $\phi_o = [-1/2,-1/2]^T$, $a = c = 0$, $b = -\rho F''(s_o)$.

Thus the conclusion of theorem 4.2 holds, namely B and C are bifurcation points, and the tangent to the bifurcating curve at B or C is given by:

$$\mu'(0) = \beta_o, \quad w'(0) = \beta_o\phi_o + \beta_1\phi$$

where (β_o,β_1) satisfies the bifurcation equation:

$$b\beta_o\beta_1 = 0.$$

The two possible solutions are (0,1) and (1,0) (up to some multiplicative factor). As $\alpha_o = \lambda'(s_o) \neq 0$, necessarily $\beta_o = 0$ and the bifurcation is one-sided.

4.4 Numerical Analysis of Bifurcations (Keller [2,3])

We have seen in Section 3.3 how to trace a smooth arc Γ of solutions for:

$$f(\lambda,u) = 0, \quad \lambda \in R^+, \ u \in R^N$$

by using a continuation method such as Kubicek method, which consists in parametrizing by the arc length s along the arc Γ

$$f(\lambda(s),u(s)) = 0 \quad \text{and} \quad \lambda'^2(s) + |u'(s)|^2 = 1,$$

and solving a system of ordinary differential equations:

$$\begin{cases} f_u(\lambda(s),u(s))u'(s) + f_\lambda(u(s),\lambda(s))\lambda'(s) = 0 \\ |\lambda'(s)|^2 + |u'(s)|^2 = 1 \end{cases}$$

with initial conditions:

$$\lambda(0) = \lambda_o, \quad u(0) = u_o$$

where (λ_o,u_o) is an isolated starting point.

Such a method, applied to problem:

$$\text{(4.41)} \quad \begin{cases} 2u_1 - u_2 + \rho F(u_1) - \lambda = 0 \\ -u_1 + 2u_2 + \rho F(u_2) - \lambda = 0, \end{cases}$$

with starting point (0,0), provides the arc OABCDZ of Figure 4.2. However the bifurcation points B and C and the bifurcating branches are ignored by such an algorithm. It is therefore desirable to enhance the method by providing the following possibilities:

(i) location of branching points
(ii) distinction of bifurcation points from turning points
(iii) computation of points on bifurcation branches in the vicinity of a bifurcation point, in order to treat them as starting points in subsequent applications of a continuation method.

These additions to the standard program of Kubicek (or any method of continuation) are easily done by following closely the results of the previous section.

(i) Determining the position of branching points: At each point $(\lambda(s),u(s))$ along the curve, we evaluate the sign of the determinant $\det(f_u(\lambda(s),u(s))$. When the determinant changes sign between successive points of the arc, we employ a bisection search to refine the position of the singular point up to an error tolerance. We suppose that this process has produced a (nearly) singular point (λ_o,u_o) corresponding to the arc coordinate $s = s_o$.

(ii) <u>Distinction of bifurcation points from turning points</u>: The point (λ_o,u_o) may possibly be either a turning point or a bifurcation point. We have seen in Section 4.3 that generically the following holds. Either:

$$\psi^* f^0_\lambda \neq 0$$

and (λ_o,u_o) is a turning point, or:

$$\psi^* f^0_\lambda = 0$$

and (λ_o,u_o) is a bifurcation point.

Figure 4.4 shows the values of $\psi^* f^0_\lambda$ at the branching points of equations (4.10), and it appears that bifurcation points have a value of $\psi^* f^0_\lambda$ which is very small $(O(10^{-10}))$, whereas for turning points $\psi^* f^0_\lambda = 1.41$. (In fact $\sqrt{2}$, as may be directly checked!)

In order to calculate $\psi^* f^0_\lambda$ we need to calculate ψ^* at the branching point (λ_o,u_o).

<u>Calculation of ϕ and ψ^*</u>: It is done most efficiently in the following way [9]. The triangularization of matrix f^{0*}_u with partial pivoting yields a lower triangular matrix L with ones on the diagonal and an upper triangular matrix U with 0 at the last diagonal term. Since L is nonsingular, solving $LU\phi = 0$ is equivalent to $U\phi = 0$, that is:

$$\begin{cases} \sum_{j=i}^{N} u_{ij}\phi_j = 0, & 1 \le i \le N-1 \\ 0\phi_N = 0 \end{cases}$$

whose solution is calculated from $\phi_N = 1$ for example by backward substitution.

For calculating ψ^* remark that if P is the permutation matrix corresponding to the partial pivoting, the solution of $f^{0*}_u\psi = 0$ satisfies $U^*L^*P\psi = 0$ and is obtained by solving successively:

$$U^*x = 0 \quad (\Longrightarrow x = (0,0,\dots,0,1)); \quad L^*y = x, \quad P\psi = y$$

(iii) Computation of points on bifurcation branches in the vicinity of a bifurcation point: Since we are currently on a particular branch Γ_1, $(\lambda'(s_o), u'(s_o))$ are known up to a multiplicative factor. We use the upper and lower bounds on (λ_o, u_o) obtained in the bisection search to estimate $(\lambda'(s_o), u'(s_o))$ (Fig. 4.7):

$$\begin{cases} \lambda'(s_o) = (\lambda_P - \lambda_M)/\varepsilon \\ u'(s_o) = (u_P - u_M)/\varepsilon \end{cases}$$

where ε is arclength between points M and P but will not appear in the final formulas.

Since $u'(s_o) = \alpha_o\phi_o + \alpha_1\phi$ with $\alpha_o = \lambda'(s_o)$ and $\psi^*\phi_o = 0$, we obtain:

$$\begin{cases} \alpha_1 = \psi^* u'(s_o)/\psi^*\phi \\ \phi_o = (u'(s_o) - \alpha_1\phi)/\lambda'(s_o). \end{cases}$$

In particular:

$$\phi_o \simeq (u_P - u_M - (\psi^*(U_P - U_M)/\psi^*\phi)\phi)/(\lambda_P - \lambda_M).$$

Now we have all the elements to calculate the coefficients a, b, c of the bifurcation equation:

$$\begin{cases} a = \psi^* f^0_{uu}\phi\phi \\ b = \psi^*(f^0_{uu}\phi_o + f^0_{u\lambda})\phi \\ c = \psi^*(f^0_{uu}\phi_o\phi_o + f^0_{u\lambda}\phi_o + f^0_{\lambda\lambda}) \end{cases}$$

The vectors $f^0_{uu}vw$ and $f^0_{u\lambda}v$ can be obtained analytically from f directly, or approximated by finite differences:

$$\begin{cases} f^0_{uu}vw \simeq \frac{1}{\varepsilon}[f_u(\lambda_o, u_o + \varepsilon w) - f_u(\lambda_o, u_o)]v \\ f^0_{u\lambda}v \simeq \frac{1}{\varepsilon}[f_\lambda(\lambda_o, u_o + \varepsilon v) - f_\lambda(\lambda_o, u_o)]. \end{cases}$$

As we already know one solution (α_o, α_1) to the bifurcation equation, the other (β_o, β_1) can be obtained simply:

$$\frac{\beta_1}{\beta_o} = -\frac{2b}{a} - \frac{\alpha_1}{\alpha_o} \quad (= r).$$

The multiplicative uncertainty in α_o, α_1 disappears in α_1/α_o . In fact:

$$\frac{\alpha_1}{\alpha_o} = \frac{\psi^*(u_P - u_M)}{\psi^*\phi(\lambda_P - \lambda_M)} .$$

Normalizing:

$$\beta_o^2(1 + |\phi_o + r\phi|^2) = 1$$

whence:

$$\beta_o = (1 + |\phi_o + r\phi|^2)^{-1/2}.$$

The tangent $(\lambda_2'(t_o), u_2'(t_o))$ to the bifurcating curve is thus determined:

$$\begin{cases} \lambda_2'(t_o) = \beta_o \\ u_2'(t_o) = \beta_o\phi_o + \beta_1\phi. \end{cases}$$

A pair of points, one on either side of the bifurcation point, and lying in the bifurcated branch are obtained by taking a short step in the direction $\pm$ $(\lambda_2'(t_o), u_2'(t_o))$. For problem (4.41) the length of this step was selected empirically to be 0.1 but it can be as small as 0.001. Evidently this estimate is subject to error, so a Newton correction procedure is applied until an error tolerance is satisfied.

<u>Determination of the connected component of solutions</u>: We are now in a position to attempt to trace out the collection of branches which are linked to an initially known solution, (λ_o, u_o). Starting from (λ_o, u_o) we trace Γ_1, the "trivial" branch, noting the occurrence of each bifurcation or turning point. As they are found, a pair of points on the putative bifurcated branch Γ_2 are obtained. The direction vectors to the bifurcated branch are simply the signs of the differences:

$$\lambda_P - \lambda_o, \quad (u_P - u_o)_i, \qquad 1 \le i \le N$$

or

$$\lambda_M - \lambda_o, \quad (u_M - u_o)_i, \qquad 1 \le i \le N,$$

where M,P denote the direction taken along Γ_2. These provide the starting points and direction vectors for the "primary" bifurcation branches. The process is continued until either the connected component is exhausted, or the state variables or the parameter λ exceed preset limits.

Application to problem (4.41) (two-cell assemblage): The case $N = 2$ with $\rho = 100$, $k = 1$ already illustrates the capacities of the algorithm and the complexity of the criteria. Figure 4.4 shows $s_1 (= u_1)$ as a function of s_o $(= \lambda)$.

There are two turning points and two bifurcation points on the trivial branch. The bifurcations are both unilateral, and lead to stable solutions after a turning point. In fact the stability of the solutions on arcs EF and GH has to be checked separately since the only information we possess a priori is that the determinant of $f_u(\lambda,u)$ changes sign as the point (λ,u) crosses, say E, so that either 2 or 0 eigenvalues are negative on arc EF.

For $N = 3$ there is another pair of bifurcation points on the trivial solution, and another closed primary bifurcation branch. There are no secondary bifurcation points, only numerous turning points, and each branch has a pair of arcs of stable solutions.

In conclusion: We have at our disposal, with the Kubicek method and Keller method for bifurcations, an efficient algorithm for the numerical exploration of a connected component of stationary solutions.

4.5 Imperfections

In equations (4.1) the substrate concentrations in both reservoirs were assumed to be the same. It is a highly idealized situation because it is practically impossible to impose exactly the same concentration in two distinct reservoirs, and the following equations, more realistic, hold instead of (4.1):

$$(4.42) \quad \begin{cases} 2s_1 - s_2 + \rho F(s_1) = s_o \\ -s_1 + 2s_2 + \rho F(s_2) = s_o + \varepsilon \end{cases}$$

where ε is supposed to be a small imperfection or lack of homogeneity.

What happens to the "trivial" solution and to the bifurcated branch of solutions in the neighborhood of a bifurcation point as the magnitude of the imperfection deviates from zero has been studied by J.P. Keener and H.B. Keller for a very broad class of problems [10]. It is the phenomenon of perturbed bifurcation. In fact, by adding a slight imperfection, the whole nature of the set of solutions is changed. The aim of this section is to show on the trivial example (4.42) what happens and try to explain it.

Figure 4.8 shows the set of solutions for (4.42) when $\varepsilon = 0$ (heavier lines), with the "trivial" branch of "symmetric" solutions OABCDZ and the bifurcated loop BEFCHGB. It shows also the set of solutions for (4.42) when $\varepsilon \neq 0$ (lighter lines), with a disconnected locus of solutions which forms an "isola" beside the rest of the set of solutions. This locus has a visual appearance which suggests, correctly, that it is composed of symmetrical and asymmetrical segments BC and BEFC. Each segment of the disconnected locus corresponds in its stability and symmetry to the analogue segment of the connected locus immediately adjacent. BC and BEFC disappear from the main locus, and only on this branch remain BGHC, OAB, and CDZ.

As ε is increased, the closed locus becomes smaller and increasingly separate from the main locus, ultimately disappearing for large values of ε [1].

<u>First explanation of perturbed bifurcation [11]</u>: On one hand we consider solutions of (4.1) such that $s_o = s + \rho F(s)$, $s_1 = s$, and $s_2 = s$, s being close to a solution b of $3 + \rho F'(b) = 0$. Thus the point of coordinates (s_o, s, s) lies on the arc AD of symmetric solutions, close to one of the two bifurcation points B and C. (Fig. 4.8). On the other hand, <u>for the same value s_o</u>, we consider nearby solutions of (4.42), (s_o, s_1, s_2). Let $u_1 = s_1 - s$ and $u_2 = s_2 - s$. Since $s + \rho F(s) = s_o$, we have:

$$2u_1 - u_2 + \rho(F(s+u_1) - F(s)) = 0$$

$$-u_1 + 2u_2 + \rho(F(s+u_2) - F(s)) = \varepsilon,$$

whence:

Application to the 2-cells assemblage with imperfection: Problem (4.42) can be rewritten $f(\lambda,\tau,u) = 0$ where $\lambda = s_o$, $\tau = \varepsilon$, $u = (u_1,u_2) = (s_1,s_2)$ and $f(\lambda,\tau,u) = (2u_1 - u_2 + \rho F(u_1) - \lambda, -u_1 + 2u_2 + \rho F(u_2) - \lambda - \tau)$. Moreover, as seen at the end of Section 4.3, $a = c = 0$ and $b = -\rho F''(b_i)$, b_i, $(i = 1,2)$, satisfying $3 + \rho F'(b_i) = 0$. Since $d = \psi^* f_\tau^0 = 1$, the bifurcation equation reduces to:

$$\tau = \rho F''(b_i)\varepsilon(\lambda-\lambda_o) + \text{h.o.t.'s.}$$

which is to be compared to (4.44).

The information brought to us by both methods is about the qualitative behavior of the solution curves near (λ_o,τ_o,u_o). In order to know their global behavior numerical methods of continuation such as those previously exposed have to be employed.

Remark 4.1 - The biochemical interpretation of the above numerical and mathematical results is that, in assemblages of cells where an enzyme reaction occurs, new non-symmetric stable steady-states may exist when the cells are separated by diffusion barriers. We shall develop this idea further in the next two chapters, where pattern formation in the so-called S-A system will result from sequential bifurcations and will be compared to morphogenesis in developing embryos.

References

[1] Bunow B. and Colton C.K., Substrate inhibition kinetics in assemblages of cells, Biosystems , 5 (1975).

[2] Keller, H.B., Numerical solution of bifurcation and nonlinear eigenvalue problems, 359-384: Rabinowitz, P.H. (ed.), Applications of Bifurcation Theory (Academic Press, New York, 1977).

[3] Keller, H.B., Constructive methods for bifurcation and nonlinear eigenvalue problems, Proceedings of a meeting in IRIA, 1-11 (1977).

[4] Keller, H.B., Global homotopies and Newton methods, 73-94: C. de Boor and G. Golub (eds.), Recent Advances in Numerical Analysis (Academic Press, New York, 1978).

[5] Decker, D.W., Topics in bifurcation theory, Ph.D. Thesis, California Institute of Technology, Pasadena, California (1978) 152 pp. + V.

[6] Szeto, R., The flow between rotating coaxial disks, Ph.D. Thesis California Institute of Technology, Pasadena, California (1978).

[7] Keener, J.P., Secondary bifurcation and multiple eigenvalues, S.I.A.M. J. Appl. Math., Vol. 7, N°2 (1979) 330-349.

[8] Crandall, M.G. and Rabinowitz, P.H., Bifurcation, perturbation of simple eigenvalues, and linearized stability, Arch. Rat. Mech. Anal. 52 (1973) 161-180.

[9] Keller, H.B., Personal communication.

[10] Keener, H.P. and Keller, H.B., Perturbed bifurcation theory, Arch. Rat. Mech. Anal., Vol. 50 (1973) 159-175.

[11] Iooss, G., Personal communication.

[12] Keller, H.B., Two new bifurcation phenomena, IRIA Research Report N°369 (1979) 1-22.

[13] Ray, W.H., Bifurcation phenomena in chemically reacting systems, 285-315: Rabinowitz, P.H. (ed.), Applications of Bifurcation Theory (Academic Press, New York, 1977).

CHAPTER 5

THE URICASE SYSTEM AND PATTERN FORMATION

Pattern formation is probably one of the most surprising phenomena which occur in reacting and diffusing systems. Its biological interest is to suggest an explanation for morphogenesis in developing organisms.

This chapter and the next one are devoted to the study of pattern formation in enzyme systems. By pattern formation we mean that stable spatially non-uniform concentration profiles supersede spatially uniform concentration profiles which become unstable as a bifurcation parameter crosses a critical value.

The systems under consideration are open systems where an external reservoir supplies a flow of substrates, the substrates diffuse with distinct diffusion coefficients and they are consumed in an enzyme reaction. In the absence of diffusion constraints in the reactive medium, an equilibrium state settles with the same concentration values in all the cells.

This so-called trivial steady state is stable, and remains stable as long as relatively small diffusion constraints are imposed to the substrates within the reactive medium. However this uniform distribution of diffusible reacting substrates loses stability as the diffusion constraints cross a critical value, and a stable patterned distribution of the substrates emerge.

This is the so-called Turing effect [1], also known as "diffusion-induced instability" or "diffusive instability".

For example Figure 5.1 shows concentration patterns corresponding to successive stable steady states, as λ varies, for the system:

$$\begin{cases} s_t - \Delta s + \lambda[R(s,a) - (s_o - s)] = 0 & \text{in } \Omega \\ a_t - \beta\Delta a + \lambda[R(s,a) - \alpha(a_o - a)] = 0 \\ R(s,a) = \rho as/(1 + s + ks^2) \\ + \text{ no-flux boundary conditions,} \end{cases}$$

where Ω is an elliptical domain:

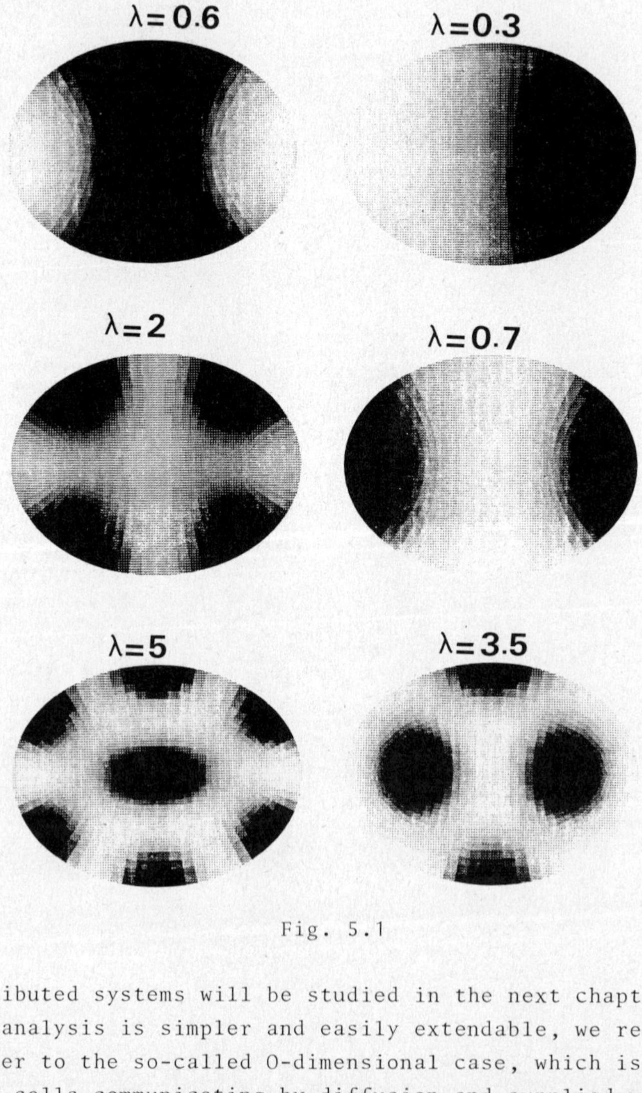

Fig. 5.1

Such distributed systems will be studied in the next chapter and, since the analysis is simpler and easily extendable, we restrict this chapter to the so-called 0-dimensional case, which is an assemblage of N cells communicating by diffusion and supplied with substrates from an external reservoir. This "zero-dimensional" system is governed by the equations:

$$(5.1)\quad \begin{cases} \dfrac{ds_1}{dt} + s_1 - s_2 + \lambda[\rho a_1 F(s_1) - (s_o - s_1)] = 0 \\ \dfrac{ds_i}{dt} - s_{i-1} + 2s_i - s_{i+1} + \lambda[\rho a_i F(s_i) - (s_o - s_i)] = 0 \quad (2 \leq i \leq N-1) \\ \dfrac{ds_N}{dt} - s_{N-1} + s_N + \lambda[\rho a_N F(s_N) - (s_o - s_N)] = 0 \end{cases}$$

and

$$(5.2)\quad \begin{cases} \dfrac{da_1}{dt} + \beta(a_1 - a_2) + \lambda[\rho a_1 F(s_1) - \alpha(a_o - a_1)] = 0 \\ \dfrac{da_i}{dt} + \beta(-a_{i-1} + 2a_i - a_{i+1}) + \lambda[\rho a_i F(s_i) - \alpha(a_o - a_i)] = 0 \\ \qquad\qquad (2 \leq i \leq N-1) \\ \dfrac{da_N}{dt} + \beta(-a_{N-1} + a_N) + \lambda[\rho a_N F(s_N) - \alpha(a_o - a_N)] = 0 \end{cases}$$

where s_i (resp. a_i) denotes the substrate S (resp. cosubstrate A) concentration in the ith cell ($1 \leq i \leq N$), s_o and a_o are the concentrations of S and A in the reservoir, α, β, ρ are positive parameters and:

$$(5.3)\quad F(s) = s/(1 + s + ks^2)$$

We explain the modeling of the N-cells assemblage in Section 5.1. In particular, we define the trivial steady-state $s_1 = \ldots = s_N = \tilde{s}$ and $a_1 = \ldots = a_N = \tilde{a}$, $\tilde{s}$ and $\tilde{a}$ being the one solution of:

$$(5.4)\quad \begin{cases} \rho\tilde{a}F(\tilde{s}) - (s_o - \tilde{s}) = 0 \\ \rho\tilde{a}F(\tilde{s}) - \alpha(a_o - \tilde{a}) = 0. \end{cases}$$

The stability of this trivial steady state is analyzed in Section 5.2. For an appropriate choice of the parameter values α, β, ρ, s_o, and a_o, the spectrum of the linearized operator consists only of real eigenvalues. We take λ as the bifurcation parameter, its physical meaning being that:

$$(5.5)\quad \lambda = \theta_D/\theta_T$$

is the ratio of θ_D and θ_T which are respectively the characteristic times of diffusion between cells and of transport of S from the reservoir to the cells. The main result of Section 5.2 is the existence of N-1 bounded intervals I_n $(1 \leq n \leq N-1)$ such that:

(i) if λ belongs to at least one of these intervals, then the trivial steady-state $(\tilde{s},\tilde{a})$ is unstable, and

(ii) if λ is exterior to all these intervals, then $(\tilde{s},\tilde{a})$ is stable.

Then we see in Section 5.3 that a stable pattern (or structure) is triggered off by random disturbances as λ crosses a critical value separating the regions of stability and instability, that is the trivial steady state transmits its stability to an emerging non-uniform steady state.

Finally the numerical techniques described in the two preceding chapters are employed in Section 5.4 to yield a complete panorama of the system behavior.

There are two reasons for studying this "zero-dimensional" case before n-dimensional cases (n = 1,2 or 3) in the next chapter. The first one is that the analysis is simpler. The second is that cell assemblages are good models for pattern formation in biology. Already in 1952 Turing [1] studied stationary waves on a ring of cells in order to explain the regeneration of the tentacles of Hydra and the appearance of the whorls of leaves of certain plants such as Woodruff.

Since then, this pioneering paper is inspiring the research of those, biologists, biochemists, or biomathematicians, who believe that the interaction of diffusion and reaction plays an important role in morphogenesis, that is the appearance of spatial structures in developing organisms.

Although in the present chapter we restrict our analysis to linear arrays of cells, the methods we employ here apply equally well to other types of networks, for example to a ring of cells as studied by Turing. For related studies in another enzyme system, we refer to [2].

5.1 Presentation of the Uricase System

5.1.1 - Case of a single cell: We first consider a single cell or compartment containing an homogeneous solution of uricase, uric acid, and oxygen. Uricase is an enzyme which catalyzes the consumption of uric acid and oxygen according to the stoichiometric scheme:

$$\text{uric acid} + \text{oxygen} \xrightarrow{\text{uricase}} \text{allantoin} + \text{other products.}$$

Uric acid and oxygen will be called respectively the substrate S and the cosubstrate A, and their concentrations will be denoted [S] and [A]. Their reaction rate was already discussed in Section 1.2 and given in formula (1.18):

$$(5.6) \qquad -\frac{d[S]}{dt} = -\frac{d[A]}{dt} = V_M \frac{[A]}{K_A} \frac{[S]}{K_S + [S](1 + [S]/K_{SS})}$$

Many reactions have similar rate expressions. For example Figure 5.2 shows the measured activity (i.e. rate expression) of the reaction:

$$\text{ATP} + \text{F6P} \xrightarrow{\text{PFK}} \text{products}$$

where PFK (phosphofructokinase) catalyzes the consumption of ATP (substrate S) and F6P (cosubstrate A). As for rate expression (5.6), we see that if [A] is held fixed the activity as a function of [S] is substrate inhibited, whereas if [S] is held fixed, it is an increasing function of [A], thus cosubstrate activated.

However we restrict our attention to the precise and simple rate expression given in (5.6). Thus the S and A concentrations in a single cell evolve according to the equations:

$$(5.7) \qquad \frac{ds}{dt} = \frac{da}{dt} = -\frac{1}{\theta_R}\, a\, F(s)$$

where:

$$F(s) = s/(1 + |s| + ks^2)$$

and:

$$s = [S]/K_S, \quad a = [A]/K_S, \quad \theta_R = K_A/V_M, \quad k = K_S/K_{SS}.$$

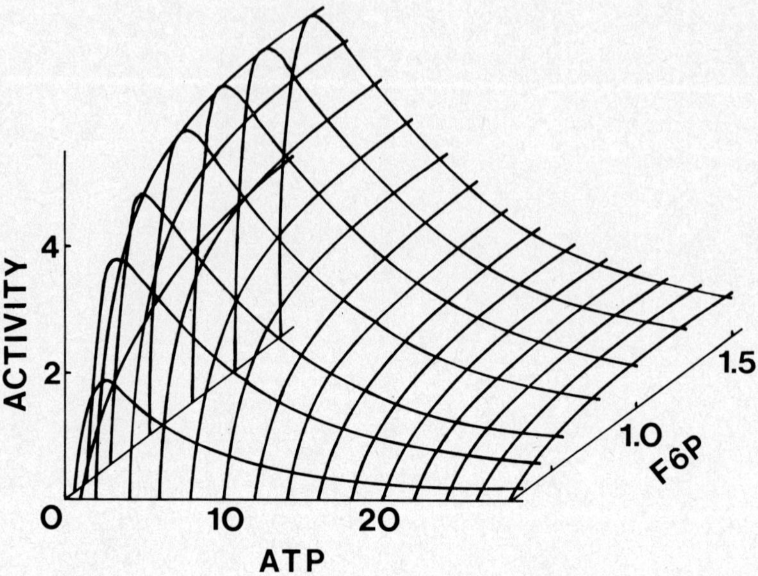

Fig. 5.2

Of the two parameters which play a role in equations (5.7), one, k, is determined by the kinetic properties of the enzyme reaction, whereas the other, θ_R, which is the reaction characteristic time, depends upon V_M which is proportional to the enzyme concentration in the cell, and thus is at our disposal.

The system represented by equations (5.7) has a rather poor behavior, both concentrations s and a converging to 0 as $t \to +\infty$.

In order to keep the activity of the system from becoming smaller and smaller, we can supply substrate and cosubstrate coming from an external reservoir through a membrane (Fig. 5.3.a). The reservoir contains a well-stirred solution of S and A at fixed concentrations s_o and a_o, and the membrane is inactive, that is without enzyme, and only plays the role of transporting S and A from the reservoir to the reactive medium with the rates:

$$\left(\frac{ds}{dt}\right)_{transport} = D'_S \frac{s_o - s}{L'} \frac{\Sigma'}{V_R} \quad \text{and}$$

$$\left(\frac{da}{dt}\right)_{transport} = D'_A \frac{a_o - a}{L'} \frac{\Sigma'}{V_R}$$

where D'_S (resp. D'_A) is the S (resp. A) diffusion coefficient in the membrane, L' the membrane thickness, Σ' the contact area and V_R the volume of the reacting solution in the cell. We define the characteristic time for transport of S from the reservoir to the cell to be:

$$\theta_T = \frac{L'V_R}{D'_S\Sigma'} = \frac{L'^2}{D'_S} \frac{V_R}{V'}$$

where V' is the membrane volume, and α to be the ratio of A and S diffusion coefficients:

$$\alpha = D'_A/D'_S.$$

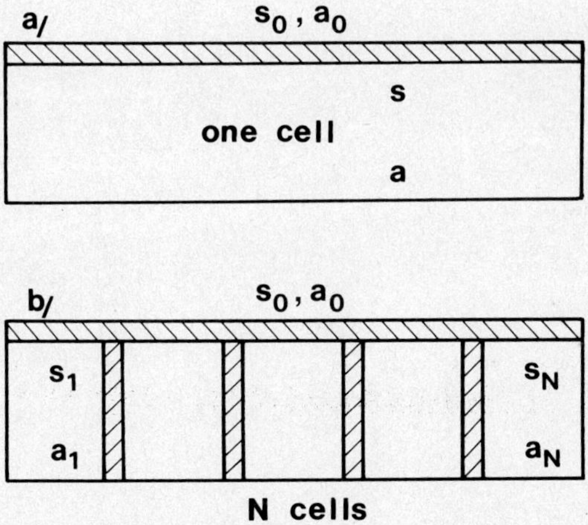

Fig. 5.3

Then the equations governing the evolution of s and a in a single cell fed from an external reservoir are:

$$(5.8) \quad \frac{ds}{dt} = \left(\frac{ds}{dt}\right)_{reaction} + \left(\frac{ds}{dt}\right)_{transport} = -\frac{1}{\theta_R} aF(s) + \frac{1}{\theta_T}(s_o - s)$$

and (5.8) continued:

$$\frac{da}{dt} = \left(\frac{da}{dt}\right)_{reaction} + \left(\frac{da}{dt}\right)_{transport} = -\frac{1}{\theta_R} aF(s) + \frac{\alpha}{\theta_T}(a_o - a).$$

5.1.2 - <u>The trivial steady state $(\tilde{s},\tilde{a})$</u>: The eventual steady states of the one-cell system (5.8) satisfy the algebraic equations:

$$(5.9) \quad \begin{cases} \rho aF(s) - (s_o - s) = 0 \\ \rho aF(s) - \alpha(a_o - a) = 0 \end{cases}$$

where:

$$(5.10) \quad \rho = \theta_T/\theta_R$$

It is fairly easy to express a as a function of s for each equation:

$$\begin{cases} a = (s_o - s)/[\rho F(s)] \\ a = \alpha a_o/[\alpha + \rho F(s)] \end{cases}$$

and to plot these two functions in the (s,a) phase-plane. Figure 5.4 shows these two curves represented respectively with dots and stars. These two curves admit at least one point of intersection, and we can choose the parameter values s_o, a_o, ρ and α so that:

(5.11) the system of equations (5.9) admits exactly one solution $(\tilde{s},\tilde{a})$ and,

(5.12) the steady-state $(\tilde{s},\tilde{a})$ is a stable node for the dynamical system (5.8)

Let:

$$(5.13) \quad R(s,a) = \rho aF(s).$$

The property that $(\tilde{s},\tilde{a})$ is a stable node is equivalent to asserting that the two eigenvalues of the Jacobian matrix:

$$(5.14)\quad \begin{bmatrix} 1+\tilde{R}_s & \tilde{R}_a \\ \tilde{R}_s & \alpha+\tilde{R}_a \end{bmatrix}$$

are strictly positive. In (5.14), $\tilde{R}_s$ and $\tilde{R}_a$ denote:

$$\tilde{R}_s = \frac{\partial R}{\partial s}(\tilde{s},\tilde{a}) = \tilde{a}F'(\tilde{s}),\ \tilde{R}_a = \frac{\partial R}{\partial a}(\tilde{s},\tilde{a}) = \rho F(\tilde{s}).$$

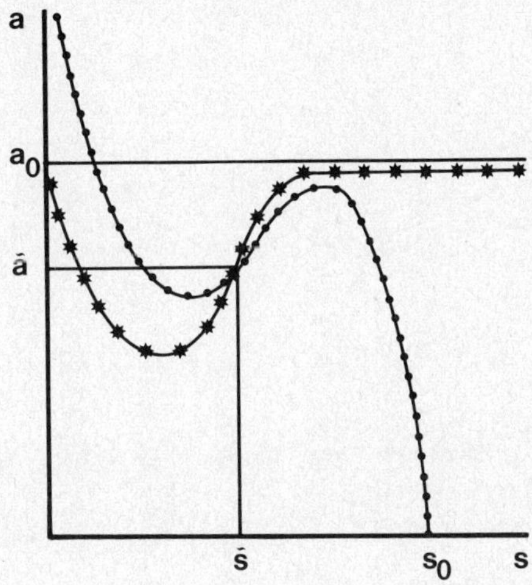

Fig. 5.4

In particular we must have, for the trace tr(O) and the determinant det(O) of the Jacobian matrix:

$$(5.15)\quad \mathrm{tr}(O) = 1+\tilde{R}_s+\alpha+\tilde{R}_a > 0 \ \text{ and } \ \det(O) = \alpha+\tilde{R}_a+\alpha\tilde{R}_s > 0.$$

Again the behavior of s and a is not very interesting; this time they converge towards $\tilde{s}$ and $\tilde{a}$ as $t \to +\infty$.

5.1.3 - N cells: However the situation becomes much more interesting when, instead of only one cell, the system consists of a linear array of N identical cells communicating by diffusion through membranes separating two neighboring cells, as pictured in Figure 5.3.b. By identical we mean that, to simplify, each cell is characterized by the same characteristic times θ_R and θ_T, is fed from the same reservoir at concentrations s_o and a_o, and is separated from its neighbors by inactive membranes having the same thickness L, the same contact area Σ, thus the same volume V, and the same diffusion coefficients D_S and D_A. Then the governing equations for the concentrations s_i and a_i in the ith cell ($2 \le i \le N-1$) are:

$$(5.16)\quad \begin{cases} \dfrac{ds_i}{dt} + \dfrac{1}{\theta_D}[-s_{i-1} + 2s_i - s_{i+1}] + \dfrac{1}{\theta_R} a_i F(s_i) - \dfrac{1}{\theta_T}(s_o - s_i) = 0 \\ \\ \dfrac{da_i}{dt} + \dfrac{\beta}{\theta_D}[-a_{i-1} + 2a_i - a_{i+1}] + \dfrac{1}{\theta_R} a_i F(s_i) - \dfrac{\alpha}{\theta_T}(a_o - a_i) = 0 \end{cases}$$

where: $\theta_D = \dfrac{LV_R}{D_S\Sigma} = \dfrac{L^2}{D_S}\dfrac{V_R}{V}$, $\beta = D_A/D_S$, and for i = 1 (resp. N) the bracketed terms have to be replaced by $s_1 - s_2$ and $a_1 - a_2$ (resp. $s_N - s_{N-1}$ and $a_N - a_{N-1}$).

Of course the system (5.16) admits the trivial solution $s_1 = s_2 = \ldots = s_N = \tilde{s}$ and $a_1 = a_2 = \ldots = a_N = \tilde{a}$, corresponding to the N cells functioning at the same levels of concentration as the previous single cell at the stable steady state $(\tilde{s},\tilde{a})$. But, even when the parameter values s_o, a_o, α and ρ are such that (5.11) and (5.12) hold, it is possible that $(\tilde{s},\tilde{a})$ be an unstable steady-state for the dynamical system (5.16). The simple fact of placing diffusion barriers destabilizes the steady-state $(\tilde{s},\tilde{a})$. In order to study more closely this phenomenon we write down the governing equations of the uricase system in the slightly more abstract setting:

$$(5.17)\quad \begin{cases} \dfrac{ds}{d\tau} + As + \lambda[R(s,a) - (s_o - s)] = 0 \\ \\ \dfrac{da}{d\tau} + \beta Aa + \lambda[R(s,a) - \alpha(a_o - a)] = 0. \end{cases}$$

where: $\tau = t/\theta_D$, $s = [s_1\ s_2\ \ldots\ s_N]^T$, $a = [a_1\ a_2\ \ldots\ a_N]^T$,

$$R(s,a) = [R(s_1,a_1)\ \ldots\ R(s_N,a_N)]^T,\quad s_o - s = [s_o - s_1\ \ldots\ s_o - s_N]^T,$$

and A is the tridiagonal matrix the non-zero terms of which are:

(5.18) $a_{11} = a_{NN} = 1, \quad a_{ii} = 2, \; 2 \leq i \leq N-1, \; a_{i,i-1} = a_{i-1,i} = -1, \quad 2 \leq i \leq N.$

For example in the case N = 5, which will be the subject of numerical simulations in Section 5.5,

(5.19) A =

1	-1	0	0	0
-1	2	-1	0	0
0	-1	2	-1	0
0	0	-1	2	-1
0	0	0	-1	+1

A is the N x N matrix of diffusion of S between neighboring cells.

The bifurcation parameter λ is the ratio:

$$\lambda = \theta_D/\theta_T$$

of the characteristic times for intercellular diffusion and reservoir-cell diffusion.

Finally the equilibrium solutions of (5.17) are given by the system of 2N nonlinear equations:

$$(5.20) \quad \begin{cases} As + \lambda[R(s,a) - (s_o - s)] = 0 \\ \beta Aa + \lambda[R(s,a) - \alpha(a_o - a)] = 0. \end{cases}$$

All the other parameters being kept fixed, we wish to study the stability of the trivial steady state $(\tilde{s},\tilde{a})$ as λ varies.

5.2 Stability of the Trivial Steady State

We are in the well known case of an ordinary differential system:

$$\frac{du}{dt} + f(\lambda,u) = 0, \quad u(t) \in R^{2N},$$

possessing a solution u_o independent of t and λ. This equilibrium solution is stable when all the eigenvalues of the Jacobian matrix $f_u(\lambda,u_o)$ have positive real part, and unstable when at least one of the eigenvalues has negative real part. The Jacobian matrix is:

$$f_u(\lambda,u_o) = \begin{bmatrix} A + \lambda(\tilde{R}_s+1) & \lambda\tilde{R}_a \\ \lambda\tilde{R}_s & \beta A + \lambda(\tilde{R}_a+\alpha) \end{bmatrix}$$

where:

$$\tilde{R}_s = \frac{\partial R}{\partial s}(\tilde{s},\tilde{a}) \quad \text{and} \quad \tilde{R}_a = \frac{\partial R}{\partial a}(\tilde{s},\tilde{a}).$$

We want to solve the eigenvalue problem:

$$f_u(\lambda,u_o)\phi = \sigma\phi.$$

This problem can be written, if $\phi = [X \;\; Y]^T$,

$$(5.21) \quad \begin{cases} AX + \lambda(\tilde{R}_s+1)X + \lambda\tilde{R}_a Y = \sigma X \\ \lambda\tilde{R}_s X + \beta AY + \lambda(\tilde{R}_a+\alpha)Y = \sigma Y. \end{cases}$$

Now we introduce the N eigenvectors w_n of A:

$$Aw_n = \mu_n w_n \qquad 0 \leq n \leq N-1.$$

Since A is symmetric the corresponding eigenvalues μ_n are real. It is obvious that $\mu_o = 0$ is an eigenvalue of A, associated with the eigenvector $w_o = [1\ 1\ \ldots\ 1]^T$. Since A is semi-positive definite the other eigenvalues w_n are strictly positive. Moreover the N eigenvectors w_n constitute an orthogonal basis of R^N. Thus X and Y admit the decomposition:

$$X = \sum_{n=0}^{N-1} \xi_n w_n \quad \text{and} \quad Y = \sum_{n=0}^{N-1} \eta_n w_n$$

and the system (5.21) can be written:

$$\sum_{n=0}^{N-1} [\mu_n\xi_n + \lambda(\tilde{R}_s+1)\xi_n - \sigma\xi_n + \lambda\tilde{R}_a\eta_n]w_n = 0$$

and

$$\sum_{n=0}^{N-1} [\lambda\tilde{R}_s\xi_n + \beta\mu_n\eta_n + \lambda(\tilde{R}_a+\alpha)\eta_n - \sigma\eta_n]w_n = 0.$$

Thus the problem is to find N pairs (ξ_n,η_n) satisfying:

$$(5.22) \quad \begin{bmatrix} \mu_n + \lambda(\tilde{R}_s+1) - \sigma & \lambda\tilde{R}_a \\ \lambda\tilde{R}_s & \beta\mu_n + \lambda(\tilde{R}_a+\alpha) - \sigma \end{bmatrix} \begin{bmatrix} \xi_n \\ \eta_n \end{bmatrix} = \begin{bmatrix} 0 \\ 0 \end{bmatrix}$$

at least one of these pairs being different from (0,0).

For (5.22) to admit a nontrivial solution we must have a vanishing determinant for the matrix of (5.22), that is σ must satisfy:

$$(5.23) \quad \sigma^2 - tr(n)\sigma + det(n) = 0$$

where:

$$\begin{cases} tr(n) = (\beta+1)\mu_n + \lambda(\tilde{R}_s + 1 + \tilde{R}_a + \alpha) \\ det(n) = \beta\mu_n^2 + \lambda[\beta(\tilde{R}_s+1) + \tilde{R}_a + \alpha]\mu_n + \lambda^2(\alpha\tilde{R}_s + \alpha + \tilde{R}_a). \end{cases}$$

Thus for each value of n $(0 \le n \le N-1)$ one obtains two eigenvalues, σ_n^- and σ_n^+, of $f_u(\lambda,u_o)$, solutions of the quadratic equation (5.23), and to each of these eigenvalues corresponds an eigenvector:

$$\phi_n^{\pm} = \begin{bmatrix} 1 \\ M_n^{\pm} \end{bmatrix} w_n$$

where $M_n^{\pm}$ is defined by the first equation of (5.22) with $\sigma = \sigma_n^{\pm}$, $\xi_n = 1$ and $\eta_n = M_n^{\pm}$:

$$\mu_n + \lambda(\tilde{R}_s+1) - \sigma_n^{\pm} + \lambda\tilde{R}_a M_n^{\pm} = 0.$$

The essential observation, due to G. Joly [3], is that, under the hypotheses:

$$(5.24) \quad 1 < \alpha < \beta$$

and $\tilde{R}_s < 0$, these eigenvalues are <u>real</u>. Indeed one verifies that:

$$\mathrm{tr}^2(n) - 4\,\det(n) = \{(\beta-1)\mu_n + \lambda[\tilde{R}_a^{1/2} - (-\tilde{R}_s)^{1/2}]^2 + \lambda(\alpha-1)\} \times$$

$$\{(\beta-1)\mu_n + \lambda[\tilde{R}_a^{1/2} + (-\tilde{R}_s)^{1/2}]^2 + \lambda(\alpha-1)\} .$$

Suppose $\sigma_n^- < \sigma_n^+$. At least one of these eigenvalues is positive since:

$$\sigma_n^- + \sigma_n^+ = \mathrm{tr}(n) = (\beta+1)\mu_n + \lambda\,\mathrm{tr}(0) > 0$$

(using the fact that $\mu_n \geq 0$ and the first inequality in (5.15)). Thus $\sigma_n^- < 0$ if and only if $\det(n) < 0$. But $\det(n) = \lambda^2 T(\mu_n/\lambda)$ where $T(z)$ is defined by:

$$T(z) = \beta z^2 + [\beta(\tilde{R}_s+1) + \tilde{R}_a + \alpha]z + \alpha\tilde{R}_s + \alpha + \tilde{R}_a .$$

Thus we are led to assume that this quadratic expression possesses two positive roots z' and z'':

$$T(z) = \beta(z-z')(z-z''), \qquad 0 < z' < z''$$

so that:

$$\det(n) < 0 \Leftrightarrow T(\mu_n/\lambda) < 0 \Leftrightarrow z' < \mu_n/\lambda < z'' \Leftrightarrow \lambda \in I_n =$$

$$]\mu_n/z'', \mu_n/z'[\ .$$

Necessary and sufficient conditions for T to admit positive roots are:

$$(5.25) \quad \begin{cases} [\tilde{R}_a + \alpha - \beta(\tilde{R}_s+1)]^2 > -4\beta\tilde{R}_s\tilde{R}_a \\ \alpha\tilde{R}_s + \tilde{R}_a + \alpha > 0 \\ \beta(\tilde{R}_s+1) + \tilde{R}_a + \alpha < 0 \end{cases}$$

We shall see below how to find parameter values in order to satisfy these conditions, but already we can see that the third condition in (5.25) implies $\tilde{R}_s < 0$.

Concluding, we have the:

Proposition 5.1 - Under the hypotheses (5.11), (5.15), (5.24), and (5.25), the trivial steady state $(\tilde{s},\tilde{a})$ is stable provided:

$$\lambda \notin \bigcup_{n=1}^{N-1} \bar{I}_n$$

and is unstable as soon as:

$$\lambda \in \bigcup_{n=1}^{N-1} I_n .$$

Choice of the parameters: Parameters k and β are given. It is not obvious that the other parameters s_o, a_o, α and ρ can be chosen so that (5.11), (5.15), (5.24), and (5.25) hold. It is possible to obtain such a set of parameters by the following procedure. (5.25) is equivalent to the fact that the polynomial T can be written:

$$T(z) = \beta(z-z')(z-z''), \qquad 0 < z' < z''$$

which itself is equivalent to:

$$\beta(\tilde{R}_s+1) + \tilde{R}_a + \alpha = -\beta(z'+z''), \; \alpha(\tilde{R}_s+1) + \tilde{R}_a = \beta z'z''$$

or

$$(5.26) \qquad \tilde{R}_s = -(z'+1)(z''+1)/(1 - \frac{\alpha}{\beta}), \quad \tilde{R}_a = (z' + \frac{\alpha}{\beta})(z'' + \frac{\alpha}{\beta})/(1 - \frac{\alpha}{\beta})$$

As a by-product we see that we must have $\alpha < \beta$ in order that $\tilde{R}_a > 0$. We know that:

$$R(s,a) = \rho a F(s) \text{ with } F(s) = s/(1 + s + ks^2).$$

Now take values for z', z'', $\frac{\alpha}{\beta}$, $\tilde{s}$ such that $0 < z' < z''$, $1 < \alpha < \beta$ and $F'(\tilde{s}) < 0$. From the second equation in (5.26) we know the value of $\tilde{R}_a = \rho F(\tilde{s})$, whence ρ, and from the first one the value of $\tilde{R}_s = \rho\tilde{a}F'(\tilde{s})$, whence $\tilde{a}$. By using equations (5.4) we obtain s_o and a_o. Then we check whether the first condition in (5.15) is satisfied and at last by considering the representative curves of:

$$a = (s_o-s)/(\rho F(s)) \quad \text{and} \quad a = \alpha a_o/(\alpha + \rho F(s))$$

we readily see whether (5.11) holds.

5.3 Bifurcation from the Trivial Steady State

Let $f : R^+ \times R^{2N} \to R^{2N}$ be the mapping:

$$f(\lambda,u) = \begin{bmatrix} As + \lambda(R(s,a) - (s_o - s)) \\ \beta Aa + \lambda(R(s,a) - \alpha(a_o - a)) \end{bmatrix}$$

where $u = \begin{bmatrix} s \\ a \end{bmatrix}$. We shall denote $u_o = \begin{bmatrix} \tilde{s} \\ \tilde{a} \end{bmatrix}$ the trivial steady state and λ_o one of the endpoints of the interval I_n $(1 \leq n \leq N-1)$: $\lambda_o = \mu_n/z''$ or μ_n/z'. We verify that (λ_o,u_o) is a bifurcation point on the branch of trivial solutions $[(\lambda,u_o),\ \lambda \in R^+]$.

(i) $f(\lambda_o,u_o) = 0$.

(ii) $N(f_u^0) = \text{span}\{\phi\}$, $N(f_u^{0*}) = \text{span}\{\psi^*\}$

where:

$$\phi = \begin{bmatrix} 1 \\ M_n^- \end{bmatrix} w_n$$

is that eigenvector of f_u associated to $\sigma_n^- = 0$ and $\psi^* = [w_n^T \quad N_n^- w_n^T]$ is the corresponding eigenvector of f_u^{0*}, defined by:

$$\begin{bmatrix} \mu_n + \lambda_o(\tilde{R}_s + 1) & \lambda_o \tilde{R}_s \\ \lambda_o \tilde{R}_a & \beta\mu_n + \lambda_o(\tilde{R}_a + \alpha) \end{bmatrix} \begin{bmatrix} 1 \\ N_n^- \end{bmatrix} = \begin{bmatrix} 0 \\ 0 \end{bmatrix} .$$

$$\phi = \begin{bmatrix} w_n \\ M_n^- w_n \end{bmatrix} \quad \text{and} \quad \psi^* = [w_n^T \quad N_n^- w_n^T] .$$

(iii) $$f_\lambda^0 = \begin{bmatrix} R(\tilde{s},\tilde{a}) - (s_o - \tilde{s}) \\ R(\tilde{s},\tilde{a}) - \alpha(a_o - \tilde{a}) \end{bmatrix} = \begin{bmatrix} 0 \\ 0 \end{bmatrix} \in R(f_u^0) .$$

(iv) The roots of the bifurcation equation are distinct.

Indeed if we parametrize the trivial branch by:

$$\lambda(t) = t, \quad u(t) = u_o$$

a first root of the bifurcation equation, (α_o, α_1), is determined by:

$$\lambda'(\lambda_o) = \alpha_o = 1, \quad u'(\lambda_o) = \alpha_o \phi_o + \alpha_1 \phi = 0$$

But $\phi_o = 0$ since $f^0_\lambda = 0$. Thus $\alpha_1 = 0$. Since α_o and α_1 must satisfy the bifurcation equation we have:

$$a\alpha_1^2 + 2b\alpha_o\alpha_1 + c\alpha_o^2 = 0$$

and we obtain c = 0, which is easily found directly, since:

$$c = \psi^*(f^0_{uu}\phi_o\phi_o + 2f^0_{u\lambda}\phi_o + f^0_{\lambda\lambda}).$$

The bifurcation equation can be written:

$$a\xi_1^2 + 2b\xi_o\xi_1 = 0$$

and the second root ($\beta_o = a$, $\beta_1 = -2b$) is distinct from (1,0) provided $b \neq 0$. But:

$$b = \psi^*(f^0_{uu}\phi_o\phi + f^0_{u\lambda}\phi) = \psi^* f^0_{u\lambda}\phi$$

and, from (5.22) with $\lambda = \lambda_o, \sigma = \sigma_{\bar{n}} = 0$, $\xi_n = 1$, and $\eta_n = M_{\bar{n}}$,

$$f^0_{u\lambda}\phi = \begin{bmatrix} \tilde{R}_s+1 & \tilde{R}_a \\ \tilde{R}_s & \tilde{R}_a+\alpha \end{bmatrix}\phi = -\frac{\mu_n}{\lambda_o}\begin{bmatrix} I & 0 \\ 0 & \beta I \end{bmatrix}\phi$$

where I is the N x N unit matrix.

Thus:

$$\psi^* f^0_{u\lambda}\phi = -\frac{\mu_n}{\lambda_o}(1 + \beta M_n^- N_n^-)\ ||w_n||^2.$$

But:

$$1 + \beta M_n^- N_n^- = 1 + \beta\,\frac{(\mu_n + \lambda_o(\tilde{R}_s+1))^2}{\lambda_o^2\tilde{R}_s\tilde{R}_a}$$

$$= 1 + \beta\,\frac{\mu_n + \lambda_o(\tilde{R}_s+1)}{\beta\mu_n + \lambda_o(\tilde{R}_a+\alpha)} = \frac{2\beta\mu_n + \lambda_o[\beta(\tilde{R}_s+1) + \tilde{R}_a + \alpha]}{\beta\mu_n + \lambda_o(\tilde{R}_a+\alpha)}.$$

Since: $\beta(\tilde{R}_s+1) + \tilde{R}_a + \alpha = -\beta(z' + z'')$,

$$\psi^* f^0_{u\lambda}\phi = ||w_n||^2 \frac{-2\beta\mu_n}{\beta\mu_n + \lambda_o(\tilde{R}_a+\alpha)} \left(\frac{\mu_n}{\lambda_o} - \frac{z' + z''}{2}\right) \neq 0$$

(in fact $\frac{\mu_n}{\lambda_o} - \frac{z' + z''}{2} = \pm \frac{z' - z''}{2}$).

Thus if the roots z' and z" of the quadratic expression T are distinct, so are the roots of the bifurcation equation.

Moreover the bifurcation will be one-sided if a = 0, which depends upon whether:

$$\sum_{i=1}^{N} w^3_{n,i} = 0$$

or not, since, as is easily checked:

$$a = \psi^* f^0_{uu}\phi\phi = \text{proportional to } \sum_{i=1}^{N} w^3_{n,i} .$$

Once we have verified that (λ_o,u_o) is a bifurcation point, we can examine further the behavior of the steady state as λ crosses λ_o and the trivial steady state u_o loses stability. For example in the case N = 5 (Figure 5.5) which will be completely explored in the next section, this occurs at points A,B,C, and H. At points A and B σ^-_1 changes sign, it is negative between A and B, and positive outside. At point C σ^-_2 changes sign, and at H σ^-_4 changes sign. But it results from the discussion of Section 4.3 that the bifurcating branches inherit the stability of the trivial branch. Moreover if the tangent at (λ_o,u_o) to the bifurcating branch is defined by:

$$\lambda'_o = \beta_o = 0, \quad u'_o = \beta_1\Phi,$$

it follows that, at least in the vicinity of the bifurcation point, $u - u_o$ is roughly proportional to:

$$\phi = \begin{bmatrix} 1 \\ \\ M^-_n \end{bmatrix} w_n.$$

Thus the vectors $s - \tilde{s}$ and $a - \tilde{a}$ are roughly proportional to w_n, that is out of all the components of the "thermal noise" which cause deviation of the state from the trivial state as the latter becomes unstable, only one is amplified, the one corresponding to w_n. The system acts as a filter which selects, in the vicinity of A and B, only the mode w_1, in the vicinity of C w_2 and in the vicinity of H w_4.

Due to the ever present small fluctuations in the concentrations of S and A in the cells, there is no chance for the trivial state $(\tilde{s},\tilde{a})$ to have a physical existence as λ crosses one of the critical values λ_o, and the fluctuations trigger off a new stable steady state characterized by unequal concentrations in the cells.

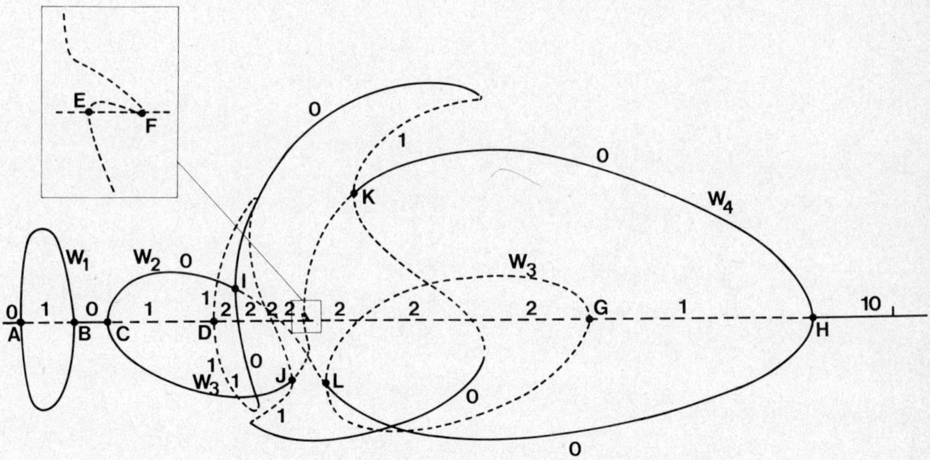

Fig. 5.5

5.4 Numerical Exploration of the Steady States [4]

It is evident that although the previous analysis gives us some information about the bifurcations from the curve of trivial steady states, however it is purely local and restricted to the behavior of the bifurcated branches near the bifurcation points. For example, we know for a given point (λ,u_o) on the trivial branch, how many negative eigenvalues lie in the spectrum of $f_u(\lambda,u_o)$. It is equal to the number of intervals I_n to which λ belongs. As this number of negative eigenvalues is transferred from the trivial branch to the bifurcating branch, we thus know the kind of stability of the steady states on that branch, at least in the vicinity of the bifurcation point. But far from the bifurcation the behavior of the branch is unknown.

It is the reason why it is necessary, if we want to know more about the whole connected set of steady state solutions to $f(\lambda,u) = 0$, to employ the numerical techniques described in the two previous chapters. Recall that all we need is a subprogram providing $f(\lambda,u)$ and $f_u(\lambda,u)$ when (λ,u) is given.

We will now give the conditions and results of the numerical experiments. We have taken $s_o = 102.5$, $a_o = 79.2$, $\alpha = 1.45$, $\beta = 5$, and $\rho = 13$. There exists a "trivial" solution for all values of λ, at, approximately, $s = \tilde{s} = 7.84$ and $a = \tilde{a} = 13.9$. This defines the trivial branch. Figure 5.5 shows for N = 5 the entire connected component issued from $(0,u_o)$, more precisely the projection of the solution onto the (λ,s_2) plane for lack of any better means of visualizing a curve in an 11-dimensional space. We hasten to point out that the solution would look quite different, had we used (λ,s_3) for example.

The horizontal coordinate is λ. The vertical coordinate is s_2. Stable solutions are shown using solid lines, while unstable solutions are indicated by dashed lines. Also indicated on the Figure are the numbers of negative eigenvalues of $f_u(\lambda,u_o)$.

Starting from $\lambda = 0$ we have indicated the trivial branch as a straight line on which appear the bifurcation points A,B,C,D,E,F,G, and H corresponding to the eigenvectors w_1 (A and B), w_2 (C and F), w_3 (D and G), and w_4 (E and H). The bifurcation points are indicated by '•'.

At each of these points there is a primary bifurcating branch of solutions. The first one has a relatively simple structure, being a closed loop passing through A and B. The bifurcation is unilateral and there is an exchange of stability. Between B and C the trivial state is stable. At C there is another bifurcation. Again the branch forms a closed loop, intersecting the trivial branch at E, F, and H.

There are 4 secondary bifurcation points, I, J, K, L on this closed loop. The bifurcation is forward at I so that the resulting bifurcation branch is stable. On the contrary, that at J is reverse, and the secondary branch is unstable. I and J are connected to distinct bifurcation branches. There is a third forward bifurcation at D. It corresponds to σ_3^- becoming negative. Since the trivial solution is already unstable with $\sigma_2^- < 0$, the bifurcating branch is unstable with one negative eigenvalue in the spectrum of the Jacobian matrix. Extending this branch in either direction there is a turning point beyond which the solutions are again stable. It is this branch at which the secondary bifurcation at I is attached. The branch emanating from J belongs to none of the primary bifurcations. Following it from J we encounter a turning point on both sides of the bifurcation at J, beyond which the solutions are stable for a considerable interval of λ. Finally, at another turning point, the branch describes only unstable solutions, closing at the tertiary bifurcation point (in this sense) K. The branch of solutions CI encounters a secondary bifurcation point L which lies on a second bifurcation branch "belonging" to w_3, that is bifurcating at G from the trivial steady state as σ_3^- crosses 0. This branch contains only unstable solutions.

The continuation of the curve CIL coincides with the branch which bifurcates at H from the trivial branch. H is one of the points on the trivial branch where σ_4^- changes sign. The other one is E but, before attaining it, the curve CILHKFE encounters first the bifurcation point K, which is a secondary bifurcation point in this sense, then the bifurcation point F. F is one of the points on the trivial branch where σ_2^- changes sign, the other being C. Thus we see that it was impossible to forecast the global behavior of the branch bifurcating from C, neither that of any bifurcating branch. However the numerical exploration of the whole connected set of solutions emanating from the trivial branch gives us a complete picture of the behavior of these solutions. Evidently, the response of the system to a variation of λ in the interval (0,10) depends upon the sense of variation. Here again if we follow by continuity the solutions of $f(\lambda,u) = 0$ as λ varies, as long as possible, then jumping when necessary from a stable branch to another one, we observe an hysteresis phenomenon when λ first increases from 0 to 10, then decreases from 10 to 0.

Lastly we can see the efficiency of the numerical method if we consider that the distance between the points E and F is of the order of 0.01, whereas the distance between points O and H is of the order of 10. The algorithm described the curve enlarged in the insert, automatically reducing the step length in the vicinity of the points E and F.

The results obtained for N = 5 and the matrix (5.19) are represented in Figure 5.5. Figure 5.6 shows the results when the matrix (5.19) is replaced by the matrix:

$$A_N = \begin{pmatrix} 2 & -2 & 0 & 0 & 0 \\ -1 & 2 & -1 & 0 & 0 \\ 0 & -1 & 2 & -1 & 0 \\ 0 & 0 & -1 & 2 & -1 \\ 0 & 0 & 0 & -2 & 2 \end{pmatrix} \qquad (5.27)$$

The advantage of matrix (5.27) is that its eigenvalues and corresponding eigenvectors are known:

$$\mu_n = 4\sin^2 \frac{n\pi}{2(N-1)}, \quad (w_n)_i = \cos n(i-1)\frac{\pi}{N-1}, \quad 1 \le i \le N, \quad 0 \le n \le N-1$$

and that the matrix:

$$(N-1)^2 A_N$$

is a finite difference approximation of $-\dfrac{d^2}{dx^2}$ with no-flux boundary conditions at $x = 0$ and $x = 1$, when the interval is divided into N-1 intervals.

<u>Remark 5.1</u> - The <u>biological interpretation</u> of the above numerical and mathematical results is that assemblages of cells, in which enzyme reaction occurs and which communicate by diffusion with one another and with an external medium, can show multiple stable steady states. There are two ways for such a system to pass from one kind

of steady state to another. Either through a stability breaking bifurcation, where the stability of the previous steady state is transferred to the new one, or at a turning point, in the neighborhood of which a slight change of an environmental parameter may cause the system to jump to an entirely different kind of steady state, thus triggering dramatic modifications.

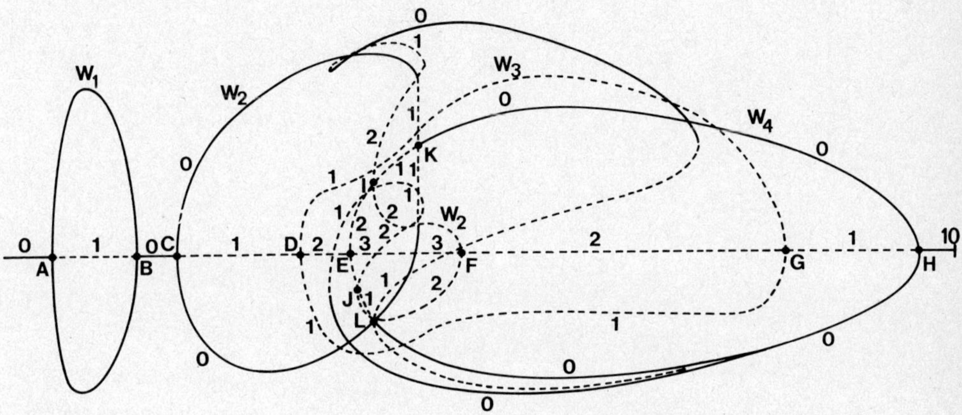

Fig. 5.6

References

[1] Turing, A.M., The chemical basis of morphogenesis, Phil. Trans. Roy. Soc. London, Vol. B 237 (1952)37-72.

[2] Bunow, B. and Colton, C.K., Multiple steady states in cellular arrays with hydrogen-ion activation kinetics, in: Thomas, D. and Kernevez, J.P. (eds.), Analysis and Control of Immobilized Enzyme Systems (North-Holland, Amsterdam, 1976).

[3] Joly, G., Thèse d'Etat, Compiègne, 1980.

[4] Bunow, B. and Kernevez, J.P., Numerical exploration of bifurcating branches of solutions to reaction-diffusion equations describing the kinetics of immobilized enzymes, (preprint).

CHAPTER 6

PATTERN FORMATION IN A MONOENZYME MEMBRANE

In this chapter we deal with pattern formation in the s-a system, governed by the equations:

$$(6.1)\quad \begin{cases} s_t - \Delta s + \lambda[R(s,a) - (s_o - s)] = 0 \quad \text{in } \Omega \\ a_t - \beta\Delta a + \lambda[R(s,a) - \alpha(a_o - a)] = 0 \\ R(s,a) = \rho a s/(1 + |s| + ks^2) \\ \text{with zero-flux boundary conditions,} \end{cases}$$

where Ω is a bounded domain in R^n ($n = 1,2$, or 3), more precisely an uricase membrane, and the conditions on the parameters s_o, a_o, α, ρ, and β are the same as in the preceding chapter. In particular the following holds:

(6.2) there is exactly one solution, $(\tilde{s},\tilde{a})$, to the algebraic system,

$$(6.3)\quad \begin{cases} R(\tilde{s},\tilde{a}) - (s_o - \tilde{s}) = 0 \\ R(\tilde{s},\tilde{a}) - \alpha(a_o - \tilde{a}) = 0 \end{cases}$$

and

(6.4) $(\tilde{s},\tilde{a})$ is a stable node for the dynamical system:

$$(6.5)\quad \begin{cases} \dfrac{ds}{dt} + R(s,a) - (s_o - s) = 0 \\ \dfrac{da}{dt} + R(s,a) - \alpha(a_o - a) = 0 \end{cases}$$

It is plain that $(\tilde{s},\tilde{a})$ is a trivial steady-state solution of (6.1). By steady state solution of (6.1) we mean a solution to the stationary problem:

$$(6.6)\quad \begin{cases} -\Delta s + \lambda[R(s,a) - (s_o - s)] = 0 \quad \text{in } \Omega \\ -\beta\Delta a + \lambda[R(s,a) - \alpha(a_o - a)] = 0 \\ \text{with zero-flux boundary conditions.} \end{cases}$$

The important fact is that, as in the preceding chapter, this equilibrium solution $(\tilde{s},\tilde{a})$ may become unstable as λ crosses a critical value. The stability analysis of $(\tilde{s},\tilde{a})$ is formally the same as in the preceding chapter, the only difference being that the states $(s(t), a(t))$ of the system are no more belonging to a finite, but an infinite dimensional vector space.

As a consequence, the spectrum of the linearized operator consists of an infinite number of eigenvalues. Spatially non-uniform steady-state solutions inherit the stability of the trivial branch as λ crosses those critical values at which an eigenvalue of this spectrum changes sign. These bifurcating solutions can be obtained by numerical methods, yielding spatially non-uniform stable steady-states even far from the bifurcation points.

The phenomena of diffusion limitation combined with the regularity properties inherent in enzymes being universal characteristics of living systems, it seems reasonable to suppose that phenomena such as pattern formation in the s-a system might well play a role in many biological processes. More precisely, while it is doubtless correct that there is a great deal more to biological pattern formation than just chemical instabilities in reaction-diffusion systems, and by no means do we suppose that uricase has any role to play in the determination of embryonic morphology in any organism, however we feel that an extensive exploration of the types of patterns to be found in the s-a system may provide a basis for studying this problem.

The origin of biological form is a problem which has challenged embryologists since the discipline was founded. Every biological organism starts life as a single cell. According to molecular biology, every detail of every phase of development is coded in the genetic material of the first cell, and embryological development procedes like clockwork. But the genetic program, by itself, is insufficient to specify all of the elements of development, and it is still a problem how the genetic information induces the formation of a pattern of differentiated cells. Some additional information is provided by the interaction of developing cells with their environment. A particular engaging suggestion is that part of this additional information is positional information, and that the information is coded in the form of concentration gradients before the actual differentiation may be observed. In a later stage, differentiation occurs according to the assigned positional information.

How does this non-homogeneous positional information is established is the essential problem. The later development is only a biochemical consequence.

One simple form of this hypothesis is very ancient in embryology: the organizing center. It is supposed that a small nucleus of cells produce a product which diffuses away, perhaps consumed by surrounding cells. The resulting field of concentration measures distance away from the organizing center. A more sophisticated version of the hypothesis was proposed by Turing [1] who pointed out that the interaction of diffusive transport and non-linear chemical kinetics of chemical substances, called morphogens, can lead to instability of their concentration field, with the result that the initially homogeneous system develops a finally non-homogeneous distribution.

Since its proposal by Turing, many workers have attempted to explain morphogenesis using reaction-diffusion models. We give, in chronological order, references to this work [2-47] which are not exhaustive. Further references can be found in the books [30, 35, 36, 43].

Recently, Kauffman, Shymko, and Trabert [40] have made a reaction-diffusion model of the sequential formation of compartmental boundaries in the embryonic differentiation of the wing imaginal disk of the fruitfly Drosophila melanogaster.

Since the s-a system (6.1) falls within the framework of their model, we thus have a strong motivation for studying pattern formation in an artificial uricase membrane, that is the succession of stable spatially non uniform steady states, solutions of (6.6), as the bifurcation parameter λ varies.

The plan of the Chapter is as follows. Pattern formation in fruitfly

Drosophila imaginal disks is explained in Section 6.1. Garcia-Bellido observations [48-51] and Kauffman's theory [40] are presented as a motivation for further study of the s-a system. The s-a system is described in Section 6.2 and its modeling yields the evolution equations (6.1) (in shortened form $du/dt + f(\lambda,u) = 0$) and the stationary equations (6.6) (in shortened form $f(\lambda,u) = 0$), a particular solution of which is a spatially uniform steady state $(\tilde{s},\tilde{a})$ satisfying (6.2), (6.3), and (6.4). The linear stability analysis of this equilibrium solution $u_o = (\tilde{s},\tilde{a})$ is performed in Section 6.3, where conditions for the eigenvalues of the Jacobian $f_u(\lambda,u_o)$ to be real are given, together with conditions for an eigenvalue to be negative.

Finally we can completely analyse the stability of the trivial steady-state u_o, mainly because this steady state is spatially uniform. Section 6.4 deals with the bifurcation analysis at those points (λ_o,u_o) where an eigenvalue of $f_u(\lambda_o,u_o)$ crosses 0. The problem is set in a more general framework than in the preceding chapters, namely the mapping:

$$f : R^+ \times X \to Y$$

involves Banach spaces X and Y. The s-a system fits this general framework and patterned solutions emanate from these bifurcation points (λ_o,u_o). How to calculate these spatially non-uniform steady states and the numerical results so obtained are indicated in Section 6.5, for the one dimensional case as well as for two-dimensional cases or diffusion reaction on a surface (involving the Laplace-Beltrami operator). Finally mathematical technical details are given in Section 6.6. In particular the spectrum of $f_u(\lambda,u_o)$ is carefully studied and a general linearized stability theorem is given, justifying the study presented in Section 6.3, and the properties of the Jacobian operator $f_u(\lambda_o,u_o)$ at the bifurcation point (λ_o,u_o) are shown justifying the analysis presented in Section 6.4.

6.1 Pattern Formation in Fruitfly Drosophila Imaginal Disks

Fruitfly Drosophila has been extensively studied by zoologists and the experimental embryology of this insect is one of the best known. The adult organs of the fly develop out of a number of approximately planar amorphous blocks of cells, called imaginal disks. There are the following pairs of disks, among others: wing, first leg, second leg, third leg, haltere, eye-antenna, and the single bilaterally fused genital disk. The disks differ from one another in their size and shape. As development procedes, the disks enlarge, while more or less retaining their original shape. We restrict ourselves to the wing disk for our explanations.

6.1.1 - Garcia-Bellido observations: The experiments of Garcia-Bellido consist in determining the ultimate position in the wing of cells in the wing disk which are marked by genetic intervention at various times during maturation. Garcia-Bellido and coworkers observed the formation of compartments, defined by the anatomic location of the progeny of early embryonic cells labelled genetically by X-irradiation. There are a series of compartmental lines which form, defined by the observation that descendants of primordial cells never cross the lines. The lines are formed sequentially, and successively subdivide the disk into progressively smaller and anatomically more specialized regions.

The five observed compartmental boundaries are shown schematically in Figure 6.1: (1) anterior-posterior: (2) dorsal-ventral : (3) wing-thorax : (4) scutum-scutellum and postscutum-postscutellum : (5) proximal-distal wing. During metamorphosis, the wing disk folds along the dorsal-ventral line (2), and apposes dorsal and ventral thorax, while the wing everts. The wing imaginal disk is not the only one on which sequential compartmental lines were observed by Garcia-Bellido and coworkers. Figure 6.2 shows the schematic compartmental lines on the leg, genital, and haltere.

6.1.2 - The model of Kauffman, Shymko, and Trabert for morphogenesis in Drosophila [40] : Kauffman, Trabert, and Shymko [40] elaborated Turing's hypothesis into a qualitative model for the detailed development of the wing imaginal disk of the fruitfly Drosophila. They proposed a model involving spontaneous development of concentration patterns from the interaction of reaction and diffusion within the disk. They observed that the position of the compartmental lines on the wing imaginal disk was reminiscent of the nodal lines of the first few eigenfunctions of the Laplacian operator on the disk surface with Neumann boundary conditions:

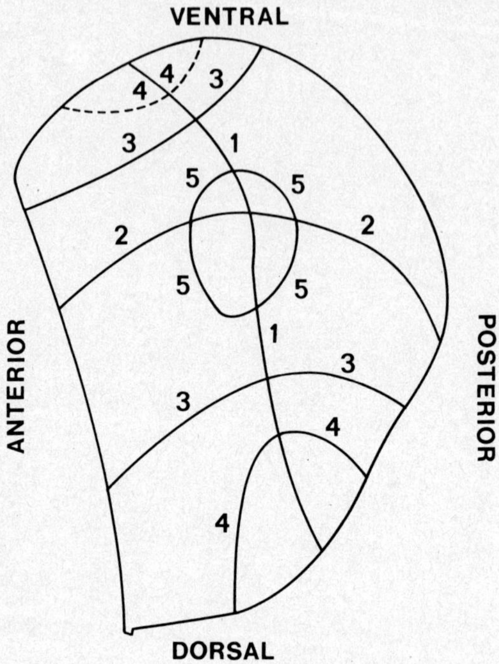
VENTRAL
4
4
3
4
3
1
5
5
2
2
5
5
ANTERIOR
1
POSTERIOR
3
3
4
4
DORSAL

Fig. 6.1

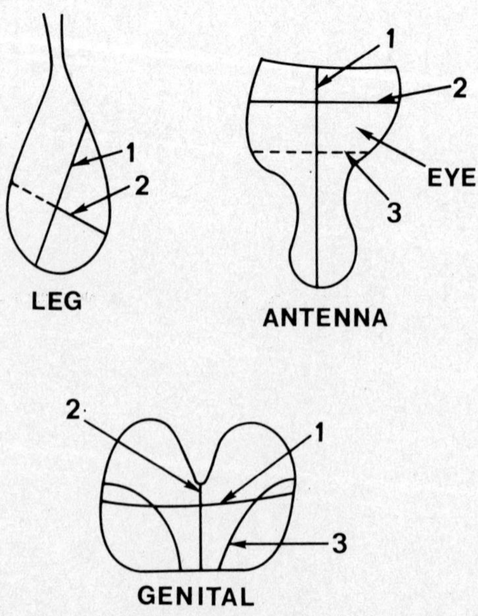
1
2
LEG
1
2
EYE
3
ANTENNA
2
1
3
GENITAL

Fig. 6.2

$$(6.7)\quad \begin{cases} -\Delta w_n = \mu_n w_n & \text{in } \Omega,\ n = 0, 1, 2, \ldots \\ \dfrac{\partial w_n}{\partial \nu} = 0 & \text{on } \partial\Omega. \end{cases}$$

Here $-\Delta$ is the Laplacian operator on the disk Ω an $\frac{\partial}{\partial \nu}$ is the (outward) normal derivative at its boundary $\partial\Omega$, μ_n are the eigenvalues, w_n the corresponding eigenvectors, $n = 0,1,2, \ldots$. For $n = 0$, $\mu_n = 0$ and $w_n \equiv 1$. The nodal lines of w_n are those lines along which $w_n = 0$. This apparently arcane similarity was the basis of suggesting that the compartmental lines were formed by cells responding to a concentration field which developed as a result of an instability in a reaction diffusion system:

$$(6.8)\quad \begin{cases} \dfrac{\partial X}{\partial t} = F(X,Y) + D_1\, \nabla^2 X \\ \dfrac{\partial Y}{\partial t} = G(X,Y) + D_2\, \nabla^2 Y \\ \text{with no flux boundary conditions} \end{cases}$$

"Such a model can have the property that its spatially uniform steady state, $X = X_o$, $Y = Y_o$ [defined by the simultaneous solution of $F(X_o,Y_o) = 0$, $G(X_o,Y_o) = 0$] is stable to all spatially distributed perturbations except those perturbations whose spatial wavelengths fall in a narrow range around some specific, characteristic wavelength, l_o."

In a linear approximation, the form of the concentration instability would resemble the eigenfunction of a Laplacian, with high concentrations on one side of the nodal line, and low on the other. Which eigenfunction is selected for amplification depends on parameter values. Since the imaginal disk is growing the parameters are changing, and it is reasonable that a sequence of successive patterns should appear. The particular sequence of patterns to be observed depends upon the shape of the domain.

In the model, as the size and shape of the disk vary , distinct chemical patterns resembling the eigenfunctions of the Laplacian emerge from the spatially homogeneous steady state and decay back to it.

Since a critical feature of compartmental boundaries is that they arise in a well defined sequence as the imaginal disk grows in size, the concentration field (X,Y) would pass, as the imaginal disk enlarges, through a sequence of stable non-uniform steady states, inducing one commitment (for example anterior) in cells where X (or Y) concentration is above some threshold level, and the alternate commitment (posterior) in cells below threshold.

This reaction-diffusion system, acting throughout development, generates a sequence of differently shaped chemical patterns.

6.1.3 - Sprey's observations [52-53]: Sprey observed the distribution of the enzyme aldehyde oxidase (AO) in the imaginal discs of some diptera, including Drosophila melanogaster. An illustration of his observations is given in Figure 6.3, which shows a clear-cut pattern of the enzyme in the left wing imaginal disc of Drosophila melanogaster. More generally Sprey observed the existence of a straight and sharp boundary between the AO-positive and AO-negative (i.e. with and without aldehyde oxidase) areas in the discs of all AO-positive flies. Interestingly, this boundary in Drosophila melanogaster coincides with a part of the borderline between the anterior and posterior developmental compartments described by Garcia-Bellido et al. [49,51] for the wing disc and by Steiner [54] for the leg disc. Also due to Sprey [55] are the photographs of Figure 6.4 and the drawings of Figure 6.5, which show actual imaginal discs. More precisely, from top to bottom, we have a Calliphora leg disc, a Calliphora wing disc and a Calliphora wing disc. It appears at once that the leg disc is not a flat disc! In fact it is a simplification to consider an imaginal disc as a planar surface. The surface of an actual imaginal disc is more complicated, involving invaginations, thick layers and thin layers of cells, etc ...

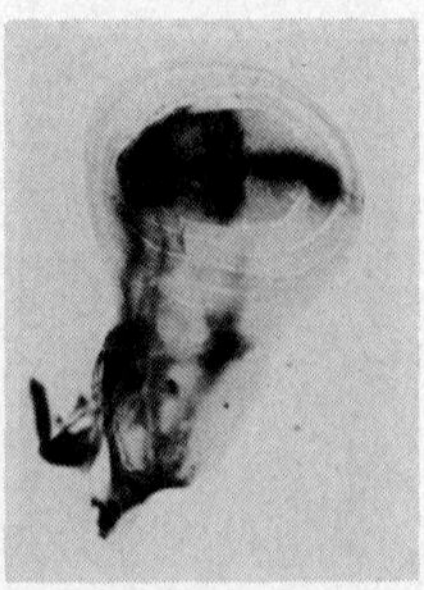

Fig. 6.3

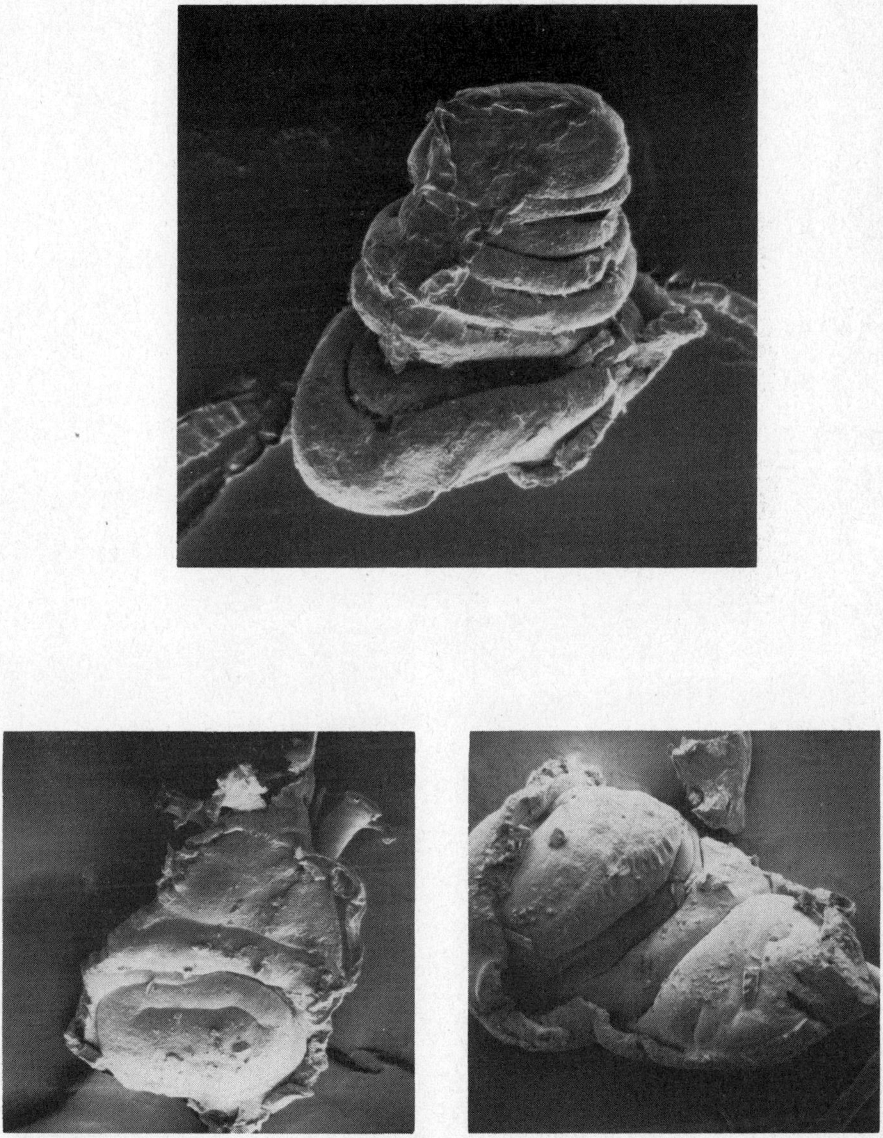

Fig. 6.4

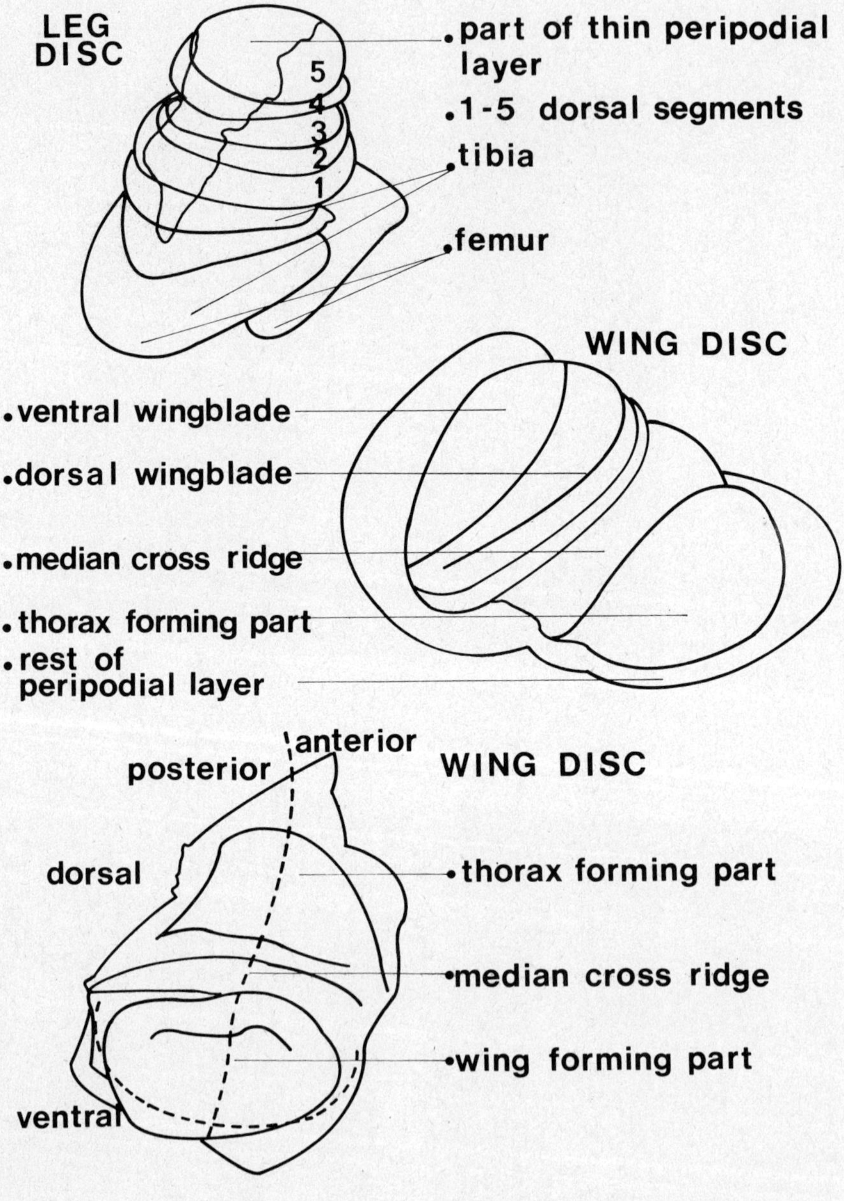
LEG DISC
5
4
3
2
1
.part of thin peripodial layer
.1-5 dorsal segments
.tibia
.femur
WING DISC
.ventral wingblade
.dorsal wingblade
.median cross ridge
.thorax forming part
.rest of peripodial layer
anterior
posterior
WING DISC
dorsal
ventral
.thorax forming part
.median cross ridge
.wing forming part

Fig. 6.5

6.2 Description of the S-A System

From the preceding section it results that reaction-diffusion could well play a non-negligible role in pattern formation. Kauffman's theory accounts for sequential compartment formation in Drosophila. Even if in the precise case of insect imaginal discs the morphogens X and Y occuring in (6.8) are hypothetic and unknown, there are various developmental systems in which the existence of morphogenetic gradients is well established experimentally [56-62]. The fact that imaginal discs are not planar surfaces is not very important since, as will result from our analysis, as soon as a given biochemical system shows pattern formation in the 0-dimensional case studied in the preceding chapter, it is able to show patterns also in various configurations, in the 1 or 2-dimensional cases, or on non planar surfaces.

Thus it is interesting to study, in a well defined context, pattern formation due to diffusive instabilities in a biochemical system. Again, artificial enzyme membranes are a good tool to achieve this goal.

6.2.1 - Modeling of the S-A system:

Figure 6.6 shows an active layer, separated from a bulk solution by an unstirred layer. e_1 and e_2 are the active and unstirred layers thicknesses. The cross section of these cylinders will not be in general circular, but will be taken to be a region shaped like the wing imaginal disc of Figure 6.1 for example. The active layer is an artificial enzymatically active membrane. The enzyme is immobilized within the membrane.

Its kinetics are those of the enzyme uricase, which catalyzes an irreversible reaction involving the two substrates oxygen and uric acid, here denoted by A and S. In an appropriate regime of concentrations, an empirical expression for the reaction rate is:

$$J_r = V_M(A/K_A)S/(K_S + S + S^2/K_{SS})$$

where S is the concentration of the substrate uric acid, and A the concentration of the co-substrate oxygen, while V_M, K_A, K_S, and K_{SS} are all constants characteristic of the enzyme defined by kinetic study in free solution.

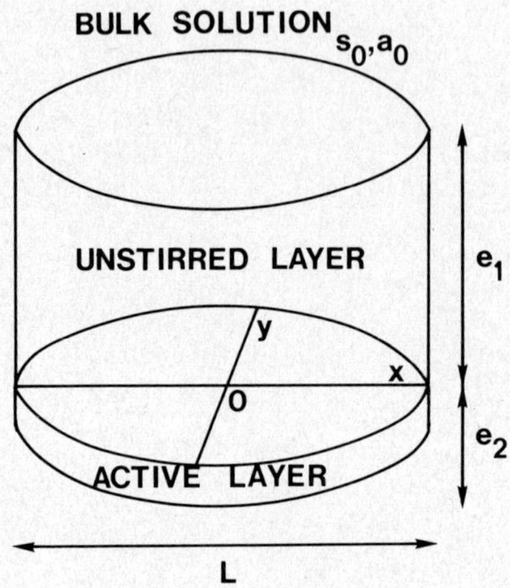

Fig. 6.6

L is some length related to the size of the membrane (diameter of its cross section for example, i.e. maximum distance between two points of this cross section). We suppose that e_1 is very small compared to L, so that transport in the narrow dimension is very rapid. Within a cross section of the active layer, however, diffusion is slow enough to permit significant concentration gradients to be established.

The modeling assumes that in the active layer the concentrations S and A depend only upon x and y, coordinates in a Cartesian frame taken in a cross section. The bulk solution is a nutrient reservoir containing the two substrates at fixed concentrations S_o and A_o. We suppose that e_2 is very small compared to L. A Fick expression for transport through the thickness of the unstirred layer is appropriate because the layer is assumed to be so thin that the concentration profile in that direction is always in a stationary state with respect to the reservoir concentrations. The equations of material balance state the condition of conservation for each of the substrates in this system:

$$(6.9)\quad \begin{cases} \dfrac{\partial S}{\partial t} = D_S\, \Delta S - J_r + D'_S(S_o - S)/(e_1 e_2) \\[2ex] \dfrac{\partial A}{\partial t} = D_A\, \Delta A - J_r + D'_A(A_o - A)/(e_1 e_2) \end{cases}$$

where D_S and D'_S (resp. D_A and D'_A) are substrate (resp. cosubstrate) diffusion coefficients in the active and inactive layers.

We now nondimensionalize (6.9) by writing:

$$(6.10)\quad \begin{cases} s = S/K_S, \quad a = A/K_S, \quad s_o = S_o/K_S, \quad a_o = A_o/K_S, \\ \theta_D = L^2/D_S, \quad \theta_T = e_1e_2/D'_S, \quad \theta_R = K_A/V_M\ , \\ \alpha = D_A/D_S, \quad \beta = D'_A/D'_S\ , \\ \rho = \theta_T/\theta_R, \ \lambda = \theta_D/\theta_T, \\ t' = t/\ \theta_D, \quad x' = x/L,\ y' = y/L. \end{cases}$$

If we substitute the non-dimensional variables t', x', y' into (6.9) and for convenience drop the primes we get:

$$(6.11)\quad \begin{cases} \dfrac{\partial s}{\partial t} - \Delta s + \lambda[R(s,a) - (s_o - s)] = 0 \\ \dfrac{\partial a}{\partial t} - \beta\Delta a + \lambda[R(s,a) - \alpha(a_o - a)] = 0 \end{cases}$$

where:

$$(6.12)\quad R(s,a) = \rho as/(1 + s + ks^2), \qquad (k = K_S/K_{SS}).$$

This biochemical system occurs in a bounded domain Ω with no flux of reactants S and A through the boundaries:

$$(6.13)\quad \frac{\partial s}{\partial \nu} = 0 \quad \text{and} \quad \frac{\partial a}{\partial \nu} = 0 \quad \text{on} \quad \partial\Omega$$

where $\frac{\partial}{\partial \nu}$ is the outward normal derivative on the boundary $\partial\Omega$ of Ω. In other words:

$$(6.14)\quad \nabla s\cdot\nu = 0 \quad \text{and} \quad \nabla a\cdot\nu = 0$$

where ∇s is the gradient of s and ν is the outward unit normal.

Note that here $0 \le x,y \le 1$ and that the domain size L has been incorporated into the parameter λ.

Thus an isotropic growth in the cross section of this system, the shape of this cross section remaining geometrically similar, will appear in our model as an increase of the parameter λ, the other parameters being unchanged and in particular the scaled domain Ω being the same.

Note also the similarity with the model of the preceding chapter. The only difference between equations (5.17) and equations (6.1) is that A has been replaced by $-\Delta$. As in Chapter 5, three characteristic times play a fundamental role: θ_D, time of diffusion within the active layer, θ_T, time of transport of S from the reservoir to the active layer, and θ_R, reaction time.

If a dimension of Ω is so narrow that transport in this dimension is very rapid, the system reduces to a one-dimensional immobilized uricase ribbon, governed by the equations:

$$(6.15)\quad \begin{cases} s_t - s_{xx} + \lambda[R(s,a) - (s_o - s)] = 0, \quad 0 < x < 1,\ t > 0, \\ a_t - \beta a_{xx} + \lambda[R(s,a) - \alpha(a_o - a)] = 0, \\ s_x(0,t) = s_x(1,t) = a_x(0,t) = a_x(1,t) = 0, \\ R(s,a) = \rho as/(1 + s + ks^2). \end{cases}$$

Such a model corresponds to situations where there exists a privileged direction.

Up to now we said nothing about the initial conditions which are necessary for the problem (6.1) to be well posed. In fact any regime of concentrations is conceivable, but we shall be mainly interested by the "trivial" (time and space independent, to be studied next) solution subject to a small, random perturbation.

6.2.2 - The spatially uniform steady-state $u_o = (\tilde{s},\tilde{a})$: Equations (6.1) always possess at least one trivial solution $s = \tilde{s}$, $a = \tilde{a}$ which is constant in both time and space. More precisely from the boundary conditions it is clear that there will exist a time and space independent solution $(\tilde{s},\tilde{a})$ if:

$$(6.16)\quad \begin{cases} s_o - \tilde{s} - \rho\tilde{a}F(\tilde{s}) = 0 \\ \alpha(a_o - \tilde{a}) - \rho\tilde{a}F(\tilde{s}) = 0 \end{cases}$$

There may be multiple solutions to these equations. We choose the parameters (s_o, a_o, α and ρ) so as to avoid this possibility, and to have a set of nulclines for the space-independent form of differential equations (6.5) similar to that shown in Figure (5.4).

Moreover we choose these parameters for $(\tilde{s},\tilde{a})$ to be a stable node for the dynamical system (6.5). A convenient set of parameters, already adopted in Chapter 5, is :

$$(6.17)\quad \begin{cases} s_o = 102.5, \quad a_o = 79.2, \ \rho = 13, \text{ and } \alpha = 1.45, \text{ for which} \\ \tilde{s} = 8 \text{ and } \tilde{a} = 14. \end{cases}$$

We have already seen in Chapter 5 that, provided $\alpha > 1$, $(\tilde{s},\tilde{a})$ is a stable node for the dynamical system (6.5) if and only if:

$$(6.18)\quad \begin{cases} tr(0) = \tilde{R}_s + 1 + \tilde{R}_a + \alpha > 0, \\ det(0) = \alpha\tilde{R}_s + \alpha + \tilde{R}_a > 0, \end{cases}$$

where $\tilde{R}_s = \frac{\partial R}{\partial s}(\tilde{s},\tilde{a}) = \rho\tilde{a}F'(\tilde{s})$ and $\tilde{R}_a = \frac{\partial R}{\partial a}(\tilde{s},\tilde{a}) = \rho F(\tilde{s})$.

6.3 Linear Stability Analysis of the Spatially Uniform Steady State $u_o = (\tilde{s},\tilde{a})$

Thus $(\tilde{s},\tilde{a})$ is a stable equilibrium point for the dynamical system (6.5). Since $(\tilde{s},\tilde{a})$ is also an equilibrium point for the dynamical system (6.1), the question is: is this equilibrium point stable ? The answer is: not always. More precisely, with an appropriate choice of the other parameters, it is possible that $(\tilde{s},\tilde{a})$ be stable or not according to the value of the parameter λ.

As in the 0-dimensional case studied in Chapter 5, diffusion can be stability breaking. In order to investigate more closely the n-dimensional case (n = 1,2 or 3) we reformulate it under the form:

$$(6.19)\quad \frac{du}{dt} + f(\lambda,u) = 0$$

where: $u = [s \quad a]^T$, $du/dt = [s_t \quad a_t]^T$ and:

$$(6.20)\quad f(\lambda,u) = \begin{bmatrix} -\Delta s + \lambda[R(s,a) - (s_o - s)] \\ \\ -\beta\Delta a + \lambda[R(s,a) - \alpha(a_o - a)] \end{bmatrix}$$

The interest of formulation (6.19) is that it is formally the same as in Chapter 5, with A instead of $-\Delta$. Evidently the essential difference is that now f is a mapping:

$$f : R^+ \times X \to Y$$

where X and Y are infinite dimensional Hilbert spaces. More precisely:

$$(6.21)\quad Y = L^2(\Omega) \times L^2(\Omega) \quad \text{and}$$

$$(6.22)\quad X = V \times V \qquad \text{where}$$

$$(6.23)\quad V = \{v \in H^2(\Omega) \mid \frac{\partial v}{\partial \nu} = 0\}.$$

However, as in the finite dimensional case, an equilibrium point u_o of (6.19) is stable if the eigenvalues of the Jacobian:

$$f_u(\lambda,u_o)$$

lie on the right-hand side of the imaginary axis in the complex plane, and is unstable if at least one of these eigenvalues lie in the left-hand side.

The reason for the stability of an equilibrium point u_o of (6.19) to be governed by the spectrum of $f_u(\lambda,u_o)$ and for this spectrum to be composed of eigenvalues will be given in Section 6.6. We restrict ourselves here to the study of these eigenvalues in the case of the trivial steady-state of the s-a system, $u_o = (\tilde{s},\tilde{a})$, which satisfies:

$$(6.24)\quad f(\lambda,u_o) = 0 \qquad \text{for every} \quad \lambda \in R^+.$$

6.3.1 - <u>Eigenvalues of $f_u(\lambda,u_o)$</u>: Eigenvalues σ and corresponding eigenvectors Φ of $f_u(\lambda,u_o)$ are non trivial solutions (σ,Φ) to the eigenproblem:

$$(6.25)\quad f_u(\lambda,u_o)\Phi = \sigma\Phi \qquad (\Phi \neq 0).$$

J.A. Boa [12,23] studied a similar problem in the case of the Bruxellator.

The procedure to study problem (6.25) will be formally the same as in the 0-dimensional case. First we define the eigenpairs (μ_n, w_n) of $-\Delta$ on Ω subject to Neumann boundary conditions on $\partial\Omega$:

$$(6.26)\quad \begin{cases} -\Delta w_n = \mu_n w_n, & n = 0, 1, 2, \dots \\ \dfrac{\partial w_n}{\partial \nu} = 0 \end{cases}$$

and we decompose:

$$\Phi = \begin{bmatrix} X \\ Y \end{bmatrix} = \begin{bmatrix} \sum_{n=0}^{\infty} \xi_n w_n \\ \sum_{n=0}^{\infty} \eta_n w_n \end{bmatrix}$$

Hereinafter we assume the w_n to be normalized:

$$\int_\Omega w_n^2(x)\,dx = 1.$$

Since:

$$f_u(\lambda, u_o)\Phi = \begin{bmatrix} -\Delta + \lambda(\tilde{R}_s+1) & \lambda\tilde{R}_a \\ \lambda\tilde{R}_s & -\beta\Delta + \lambda(\tilde{R}_a+\alpha) \end{bmatrix} \begin{bmatrix} X \\ Y \end{bmatrix}$$

$$= \begin{bmatrix} \sum_{n=0}^{\infty} \{[\mu_n + \lambda(\tilde{R}_s+1)]\xi_n + \lambda\tilde{R}_a\eta_n\}w_n \\ \sum_{n=0}^{\infty} \{\lambda\tilde{R}_s\xi_n + [\beta\mu_n + \lambda(\tilde{R}_a+\alpha)]\eta_n\}w_n \end{bmatrix},$$

Φ is an eigenvector associated to the eigenvalue σ if and only if there is, for at least one value of n, a non trivial solution to:

$$\begin{bmatrix} \mu_n + \lambda(\tilde{R}_s+1) - \sigma & \lambda\tilde{R}_a \\ \lambda\tilde{R}_s & \beta\mu_n + \lambda(\tilde{R}_a+\alpha) - \sigma \end{bmatrix} \begin{bmatrix} \xi_n \\ \eta_n \end{bmatrix} = \begin{bmatrix} 0 \\ 0 \end{bmatrix}.$$

Therefore, exactly like in the 0-dimensional case, to each eigenvalue μ_n of $-\Delta$ correspond two eigenvalues σ_n^- and σ_n^+ of $f_u(\lambda,0)$. They are the roots of the dispersion equation:

$$(6.27) \quad \sigma^2 - tr(n)\sigma + det(n) = 0$$

where tr(n) and det(n) are given in Section 5.2. Exactly like in the preceding chapter, these roots are real, provided $\tilde{R}_s < 0$, $\alpha > 1$, $\beta > 1$. Let $\sigma_n^- < \sigma_n^+$. Associated eigenvectors ϕ_n^- and ϕ_n^+ are:

$$(6.28) \quad \phi_n^\pm = \begin{bmatrix} 1 \\ M_n^\pm \end{bmatrix} w_n$$

where $M_n^\pm$ is defined by:

$$(6.29) \quad \mu_n + \lambda(\tilde{R}_s+1) - \sigma_n^\pm + \lambda\tilde{R}_a M_n^\pm = 0.$$

Moreover, since $\lambda_n^- + \lambda_n^+ = tr(n) > 0$, at least $\lambda_n^+ > 0$.

6.3.2 - <u>Conditions for σ_n^- to be negative</u>: Following the same lines as in Chapter 5, we arrive at the conclusion that:

$$\sigma_n^- < 0 \Leftrightarrow \lambda \in I_n =]\mu_n/z'', \mu_n/z'[\ , \quad n = 0, 1, 2, \ldots$$

where z' and z" are the positive roots of the quadratic expression T defined in Section 5.2. Conditions for T to admit two positive roots are expressed in (5.25).

6.3.3 - <u>Conclusion</u>: Under the hypotheses (6.2), (6.18), (5.24), and (5.25), $(\tilde{s},\tilde{a})$ is a stable steady state of the dynamical system (6.1) provided λ does not belong to any of the closed intervals $\overline{I}_n$, and is unstable if λ belongs to at least one of these intervals.

It is worth pointing out that, in the one-dimensional case for example, we have:

$$(6.30) \quad \mu_n = n^2\pi^2, \quad w_n = \cos n\pi x$$

and that, for n large enough, the intervals I_n overlap. More precisely, as long as $\mu_n/z' < \mu_{n+1}/z''$, there is, between I_n and I_{n+1}, an interval of stability for $(\tilde{s},\tilde{a})$. This condition can be written:

(6.31) $n < 1/((z''/z')^{1/2} - 1)$.

But, for n larger than this value, the intervals I_n overlap. This is shown in Figure 6.7, where interval I_n is denoted $]c_n, d_n[$, and where it can be seen that for $n \geq 2$, the intervals I_n overlap.

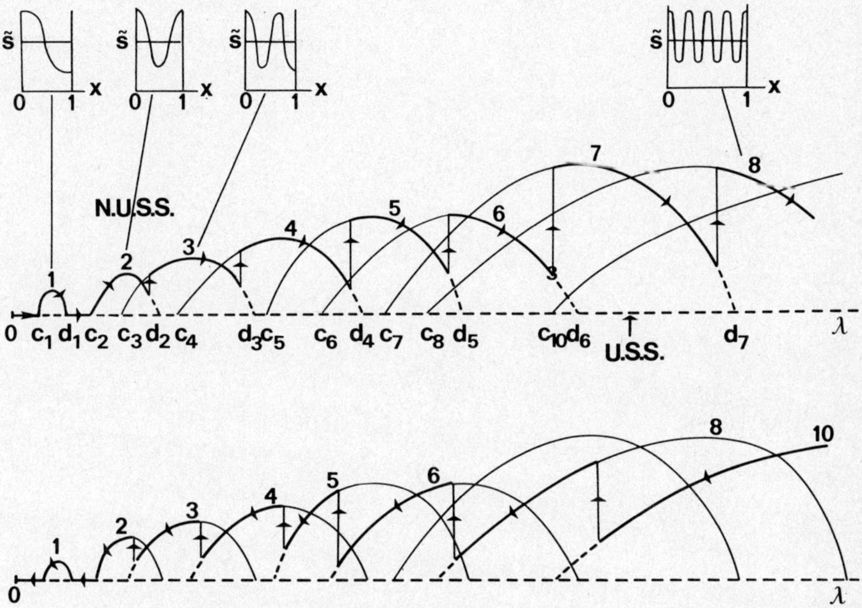

Fig. 6.7

Figure 6.7 also shows that at the points c_n and d_n spatially non uniform solutions bifurcate from the trivial branch, represented by a straight line. These bifurcated profiles are characterized by n nodes (i.e. the profile of s intersects n times the horizontal $s = \tilde{s}$) if they bifurcate at c_n or d_n $(n = 1,2,\ldots)$. The object of the next section is to explain this kind of behavior.

Finally note that the reason for the linearized stability analysis to be so simple is that we linearize about a space independent steady state. Thus the partial derivatives $\tilde{R}_s$ and $\tilde{R}_a$ are constant, and the eigenvalues and eigenvectors of $f_u(\lambda, u_o)$ are easily related to those of $-\Delta$.

6.4 Bifurcation Analysis

Before showing why bifurcation occurs in the s-a system we present a general framework within which it is practical to deal with bifurcation, at both the mathematical and numerical points of view. This framework is due to H.B. Keller [63,64] and described by D.W. Decker in [65]. Then we explain how this structure fits for the s-a system and indicate why bifurcation occurs from the "trivial" branch of the s-a system as λ crosses the endpoints of intervals I_n. We do not intend to enter into mathematical details in this justification, and delay technicalities until Section 6.6.

6.4.1 - General framework: Consider the equation:

(6.32) $f(\lambda,u) = 0$

where f is a twice continuously Fréchet differentiable mapping of $R^+ \times X$ into Y. Here X and Y are real Banach spaces, with $X \subset Y$, the embedding being continuous. We suppose that (λ_o,u_o) is a solution of (6.32) possessing the following properties:

(i) the Fréchet derivative $f_u^0 = f_u(\lambda_o,u_o)$ and its adjoint f_u^{0*} have a one-dimensional null space:

a) $N(f_u^0) = \text{span}\,\{\Phi\}, \quad \Phi \in X;$

(6.33)

b) $N(f_u^{0*}) = \text{span}\,\{\psi^*\}, \quad \psi^* \in Y^*.$

Here Y^* denotes the dual space of Y.

Further we assume the structure of the zero eigenvalue of f_u^0 to be such that:

(6.33) c) $\psi^*\phi \neq 0$

This is equivalent to demanding that this eigenvalue be simple, that is:

$\phi \notin R(f_u^0).$

(ii) We require the range of f_u^0 to be closed and of codimension 1:

(6.33) d) $R(f_u^0)$ is closed, codim $(R(f_u^0)) = 1$.

It results from the fact that $R(f_u^0)$ is closed that:

(6.34) a) $R(f_u^0) = Y_1 = \{y \in Y : \psi^* y = 0\}$

Thus:

(6.34) b) $Y = \text{span}\,\{\phi\} \oplus Y_1$.

On the other hand we have the decomposition:

(6.34) c) $X = \text{span}\,\{\phi\} \oplus X_1$

where:

(6.34) d) $X_1 = \{x \in X : \psi^* x = 0\}$.

(iii) Another requirement is that:

(6.35) $\psi^* f_\lambda^0 = 0$, i.e. $f_\lambda^0 \in R(f_u^0)$.

As a consequence there exists an unique $\Phi_o \in X$ such that:

(6.36) $f_u^0 \Phi_o + f_\lambda^0 = 0, \quad \psi^* \Phi_o = 0$.

(iv) Finally we assume the roots of the bifurcation equation to be distinct. More precisely, we recall that the bifurcation equation is:

(6.37) $a\xi_1^2 + 2b\xi_o\xi_1 + c\xi_o^2 = 0$

where:

$$(6.38) \begin{cases} a = \psi^* f_{uu}^0 \phi\phi \\ b = \psi^* (f_{u\lambda}^0 + f_{uu}^0 \phi_o)\phi \\ c = \psi^* (f_{\lambda\lambda}^0 + 2f_{u\lambda}^0 \phi_o + f_{uu}^0 \phi_o \phi_o) \end{cases}$$

and that the root (ξ_o,ξ_1) is said to be <u>isolated</u> if:

$$(6.39)\quad \begin{cases} a\xi_1 + b\xi_o \neq 0 & \text{if} \quad \xi_o \neq 0 \\ b \neq 0 & \text{if} \quad \xi_o = 0. \end{cases}$$

It is easy to check the equivalence between the existence of distinct roots (i.e. distinct rays $[(k\xi_o,k\xi_1);\ k \in R]$) and the property to be isolated for any of these roots.

Then we have the following theorem, which is a particular case of a more general result proven by Decker [65] and Keller [64] in the case of multiple bifurcation (i.e. when $\dim N(f_u^0) = \operatorname{codim} R(f_u^0) = m \geq 1$).

<u>Theorem 6.1 (Keller and Decker)</u> - Let $f(\lambda,u)$ be a C^k map $(k \geq 2)$ from $R \times X$ into Y. Let (λ_o,u_o) be a solution of (6.32) for which f_u^0 satisfies (6.33) and $f_\lambda^0 \in R(f_u^0)$. Let (ξ_o^0,ξ^0) be an isolated root of the Bifurcation Equation,(6.37). Let v^0 be the unique solution of:

$$(6.40)\quad \begin{cases} \text{a)}\ f_u^0 v^0 = -[f_{uu}^0 \Phi^0\Phi^0 + 2f_{u\lambda}^0 \Phi^0 \xi_o^0 + f_{\lambda\lambda}^0 \xi_o^{0\,2}], \\ \text{b)}\ v^0 \in X_1 \end{cases}$$

where:

$$(6.40)\quad \text{c)}\ \phi^0 \equiv \xi_o^0 \phi_o + \xi^0 \phi$$

and Φ_o is defined in (6.36).

Then for some $\varepsilon_o > 0$ there exist unique C^{k-2} smooth functions $[\xi(\varepsilon),\xi_o(\varepsilon),v(\varepsilon)]$ such that for all $|\varepsilon| < \varepsilon_o$:

$$f(\lambda(\varepsilon),u(\varepsilon)) = 0$$

where:

$$(6.41)\quad \begin{cases} \text{a)}\ u(\varepsilon) \equiv u_o + \varepsilon r(\varepsilon) + \frac{\varepsilon^2}{2} v(\varepsilon),\ \lambda(\varepsilon) = \lambda_o + \varepsilon\xi_o(\varepsilon); \\ \text{b)}\ r(\varepsilon) \equiv \xi_o(\varepsilon)\Phi_o + \xi(\varepsilon)\Phi,\quad v(\varepsilon) \in X_1; \\ \text{c)}\ \xi(0) = \xi^0,\quad \xi_o(0) = \xi_o^0, v(0) = v^0. \end{cases}$$

Note that the geometric meaning of ξ_o and ξ is that:

(6.42) $\lambda'(0) = \xi_o$ and $u'(0) = \xi_o\Phi_o + \xi\Phi$.

6.4.2 - Application to the s-a system: We apply theorem 6.1 to the mapping f defined in (6.20) and the pair (λ_o,u_o) where $u_o = (\tilde{s},\tilde{a})$ is the trivial steady state of the s-a system and λ_o an endpoint (c_n or d_n) of an interval I_n (n = 1,2,...). We review the hypotheses of this theorem.

The first one, about the C^2 smoothness of f in both arguments, will be verified in Section 6.6.

Property (6.33)a) is the very definition of λ_o. More precisely, for $\lambda = \lambda_o$ there is an integer $n \in \{1,2,...\}$ such that $\sigma_n^- = 0$, and the associated eigenvector $\phi_n^- = \phi$, defined by:

(6.43) $\Phi = \begin{bmatrix} 1 \\ M_n^- \end{bmatrix} w_n, \quad \mu_n + \lambda_o(\tilde{R}_s+1) + \lambda_o\tilde{R}_a M_n^- = 0,$

spans $N(f_u^0)$.

Property (6.33) b): Since:

$$f_u^*(\lambda,u_o) = \begin{bmatrix} -\Delta + \lambda(\tilde{R}_s+1) & \lambda\tilde{R}_s \\ \lambda\tilde{R}_a & -\beta\Delta + \lambda(\tilde{R}_a+\alpha) \end{bmatrix}$$

each eigenvalue $\sigma_n^{\pm}$ of $f_u(\lambda,u_o)$ is also an eigenvalue of $f_u^*(\lambda,u_o)$, with the associated eigenvector:

(6.44) $\psi_n^{\pm} = \begin{bmatrix} 1 \\ N_n^{\pm} \end{bmatrix} w_n$

where $N_n^{\pm}$ is defined by:

(6.45) $\begin{bmatrix} \mu_n + \lambda(\tilde{R}_s+1) - \sigma_n^{\pm} & \lambda\tilde{R}_s \\ \lambda\tilde{R}_a & \beta\mu_n + \lambda(\tilde{R}_a+\alpha) - \sigma_n^{\pm} \end{bmatrix} \begin{bmatrix} 1 \\ N_n^{\pm} \end{bmatrix} = \begin{bmatrix} 0 \\ 0 \end{bmatrix}.$

In particular for $\lambda = \lambda_o$, $\sigma_n^- = 0$ and $N(f_u^{0*}) = \text{span}\,\{\psi\}$ where:

$$(6.46)\quad \psi = \begin{bmatrix} 1 \\ N_n^- \end{bmatrix} w_n, \quad \mu_n + \lambda_o(\tilde{R}_s+1) + \lambda_o \tilde{R}_s N_n^- = 0.$$

Property (6.33) c): Since $Y = L^2(\Omega) \times L^2(\Omega)$, Y^* can be identified with Y and, since the scalar product of two elements $u = (u_1, v_1)$ and $v = (v_1, v_2)$ of Y is:

$$(u,v) = \int_\Omega [u_1(x)v_1(x) + u_2(x)v_2(x)]dx$$

we have:

$$\psi^*\phi = \int_\Omega (1 + M_n^- N_n^-) w_n^2(x)dx = 1 + M_n^- N_n^- .$$

From the definitions of M_n^- in (6.43) and N_n^- in (6.46) it results that $\tilde{R}_a M_n^- = \tilde{R}_s N_n^-$, so that:

$$1 + M_n^- N_n^- = 1 + M_n^{-2} \frac{\tilde{R}_a}{\tilde{R}_s} = 1 + M_n^{-2} \frac{\lambda_o^2 \tilde{R}_a^2}{\lambda_o^2 \tilde{R}_s \tilde{R}_a}$$

$$\overset{(1)}{=} 1 + M_n^{-2} \frac{\lambda_o^2 \tilde{R}_a^2}{[\mu_n + \lambda_o(\tilde{R}_s+1)][\beta\mu_n + \lambda_o(\tilde{R}_a+\alpha)]} \overset{(2)}{=} 1 + \frac{\mu_n + \lambda_o(\tilde{R}_s+1)}{\beta\mu_n + \lambda_o(\tilde{R}_a+\alpha)}$$

$$= \frac{(\beta+1)\mu_n + \lambda_o(\tilde{R}_s+1+\tilde{R}_a+\alpha)}{\beta\mu_n + \lambda_o(\tilde{R}_a+\alpha)} \overset{(3)}{=} \frac{tr(n)}{\beta\mu_n + \lambda_o(\tilde{R}_a+\alpha)} \overset{(4)}{=} \frac{\sigma_n^+}{\beta\mu_n + \lambda_o(\tilde{R}_a+\alpha)}$$

$$> 0.$$

Equality (1) uses the fact that the determinant of the matrix:

$$\begin{bmatrix} \mu_n + \lambda(\tilde{R}_s+1) & \lambda\tilde{R}_a \\ \lambda\tilde{R}_s & \beta\mu_n + (\tilde{R}_a+\alpha) \end{bmatrix}$$

vanishes, equality (2) uses the definition of M_n^- in (6.43), and equalities (3) and (4) come from the definition of tr(n) and the fact that $tr(n) = \sigma_n^+ + \sigma_n^- = \sigma_n^+$.

Thus zero is a simple eigenvalue of f_u^0.

Property (6.33) d) will be proven in Section 6.6.

Property (6.35) results at once from the fact that:

$$f_\lambda^0 = \begin{bmatrix} R(\tilde{s},\tilde{a}) - (\tilde{s}-s_o) \\ \\ R(\tilde{s},\tilde{a}) - \alpha(\tilde{a}-a_o) \end{bmatrix} = \begin{bmatrix} 0 \\ \\ 0 \end{bmatrix}$$

and implies that the vector ϕ_o defined by (6.36) is 0.

(6.47) $\phi_o = 0.$

Isolation of the roots of the bifurcation equation: As a consequence the coefficients of the bifurcation equation are:

(6.48) $a = \psi^* f_{uu}^0 \phi\phi$, $b = \psi^* f_{u\lambda}^0 \phi$ and $c = 0.$

Thus the bifurcation equation is:

(6.49) $a\xi_1^2 + 2b\xi_o\xi_1 = 0$

a root of which is $\xi_o = 1$, $\xi_1 = 0$. Since $\xi_o \neq 0$, the condition for this root to be isolated is $a\xi_1 + b\xi_o = b \neq 0$, which is realized since, like in Chapter 5, it can be shown that:

$$b = (\int_\Omega w_n^2 dx) \frac{\pm\beta\mu_n(z'-z'')}{\beta\mu_n + \lambda_o(\tilde{R}_a+\alpha)}$$

where z' and z" are the (distinct) roots of the quadratic expression T defined in Section 5.2.

This first root $[\xi_o = 1, \xi_1 = 0]$ clearly corresponds to the trivial branch parametrized by λ.

Thus the second root $[\xi_o = a, \xi_1 = -2b]$ corresponds to a bifurcating branch. More precisely:

Proposition 6.1: There exist unique C^0 smooth functions $(\xi(\varepsilon), \xi_o(\varepsilon), v(\varepsilon))$ such that for all $|\varepsilon| < \varepsilon_o$:

$$f(\lambda(\varepsilon),u(\varepsilon)) = 0,$$

where:

(6.50) a) $u(\varepsilon) = u_o + \varepsilon\xi(\varepsilon)\phi + \frac{\varepsilon^2}{2} v(\varepsilon); \ \lambda(\varepsilon) = \lambda_o + \varepsilon\xi_o(\varepsilon);$

and:

(6.50) b) $\xi(0) = -2b, \ \xi_o(0) = a, \ v(0) = 0.$

Proof: All the hypotheses of Theorem 6.1 are verified, except the C^2 smoothness of f and the properties that the range of f^0_u is closed and of codimension 1, which are verified in Section 6.6. Moreover, since $\phi_o = 0$ and $f^0_{\lambda\lambda} = 0$, we have:

(6.51) $v^0 = 0.$

It is easy to verify that, if $u^* = (u_1,u_2) \in Y^*$, $v = (v_1,v_2)^T \in X$ and $w = (w_1,w_2)^T \in X$, then $f^0_{uu}vw = (\tilde{R}_{ss}v_1w_1 + \tilde{R}_{sa}(v_1w_2 + v_2w_1) + \tilde{R}_{aa}v_2w_2)\begin{bmatrix}1\\1\end{bmatrix}$ so that:

$$u^*f^0_{uu}vw = \int_\Omega (\tilde{R}_{ss}v_1w_1 + \ldots)(u_1 + u_2)dx.$$

Therefore:

$$a = \psi^*f^0_{uu}\phi\phi = [\int_\Omega w_n^3(x)dx](1 + N_n^-)(\tilde{R}_{ss} + 2M_n^-\tilde{R}_{sa} + M_n^{-2}\tilde{R}_{aa})$$

and the bifurcation is one sided (i.e. $\lambda'(0) = a = 0$) if:

(6.52) $\int_\Omega w_n^3(x)dx = 0 .$

In the case of a 2- or 3-dimensional domain Ω with arbitrary shape, there is no reason for (6.52) to hold. However, in the 1-dimensional case (6.15), where (see (6.30)) $w_n(x) = \cos n\pi x$,

$$\int_0^1 w_n^3(x)dx = 0.$$

Also when Ω has an axis of symmetry, say $0 - y$ in the 2-dimensional case, suppose w is an eigenfunction:

$$-\Delta w(x,y) = \mu w(x,y), \quad \frac{\partial w}{\partial \nu} = 0, \quad \int_\Omega w^2 dx = 1,$$

and consider $v(x,y) = w(-x,y)$:

$$-\Delta v(x,y) = -\Delta w(-x,y) = \lambda w(-x,y) = \lambda v(x,y),$$

which, together with the boundary condition and the normalization:

$$\int_\Omega v^2 dx = 1,$$

implies that $v(x,y) = \pm w(x,y)$, i.e. $w(-x,y) = \pm w(x,y)$, so that there is the possibility for w to satisfy $w(-x,y) = -w(x,y)$. In that case:

$$\int_\Omega w^3 dx = 0$$

and (6.52) holds.

The case $\Omega =]0,1[$ is not unrealistic. It corresponds to the modeling of an ellipsoid with large axis ratio when one is interested in the first eigenmodes. One can thus explain the differentiation of regions in an insect egg by a pattern of concentrations in the privileged direction [40]. More generally, the onset of polarity [66] can be explained this way. The establishment of a pattern of the form:

$$s(x) = \tilde{s} + \varepsilon \cos \pi x + O(\varepsilon^2) \quad 0 < x < 1$$

$$a(x) = \tilde{a} + \varepsilon M_1^- \cos \pi x + O(\varepsilon^2)$$

which corresponds, according to (6.50), to a value of λ in the vicinity of $\lambda_o = c_1$ or d_1, has been invoked to explain head-tail distinction.

Of course a one-sided bifurcation at (c_1, u_o) for example, gives rise to a stable pattern only if it is forward. If it is reverse the bifurcating states will be unstable, at least until a turning point is encountered. In order to see whether the bifurcation is forward or reverse we need the sign of $\lambda''(0)$. But calculation of $\lambda''(0)$ is tedious and the result complicated. Thus we are one more time led to the conclusion that the best way to see what actually occurs is to employ numerical methods.

6.5 Numerical Methods and Results

We now propose to obtain the patterns which appear sequentially as λ increases. Recall that increasing λ is equivalent to growth without deformation of the domain. Two kinds of numerical methods are at our disposal for obtaining the solutions of (6.6) as λ varies.

(i) numerical methods for continuation and bifurcation of solutions as already employed in the 0-dimensional case in Chapter 5; they apply to finite difference or finite element discretizations of problem (6.6). Unfortunately these methods become very time consuming as the number of degrees of freedom increases.

(ii) methods in which the steady states are obtained by solving the evolution equations (6.1) until one estimates that the system has attained a stable steady state. However these methods do not permit obtaining unstable steady-states.

Since, anyway, our aim is to obtain sequential stable steady states of (6.1) as λ increases, we will describe methods of the second type and their results.

Since monotonically increasing values of λ correspond to a spatial domain enlarging and remaining geometrically similar to the shape of Ω, a numerical experiment appropriate for following pattern formation in the s-a system is the following:

We find the numerical solution of the initial boundary value problem (6.1) for a succession of values λ and use at each step the steady state of the preceding step as an initial condition. If the step in λ is small, this procedure is analogous to following the evolution of the solution on a growing domain.

Everytime λ is changed, the preceding steady-state is triggered off by random disturbances in order to have the whole range of spatial wavelengths in the Fourier expansion of the initial condition. In fact these environmental conditions, so that the system may undergo its sequential alterations, seem to be a good simulation of the ever present "thermal" noise in Nature.

All the numerical experiments to be described hereinafter were performed with the following values of the parameters:

(6.53) $s_o = 102.5$, $a_o = 79.5$, $\alpha = 1.45$, $\rho = 13$, $k = 0.1$, and $\beta = 5$,

satisfying the conditions (6.2), (6.18), (5.24), and (5.25) for pattern formation to occur.

We (i.e. B. Bunow, M.C. Duban, G. Joly, and the Author) used the finite difference method in the 1-dimensional case, and the finite element method in the 2-dimensional case (either on a planar disc or on a more general surface).

We discretize with respect to time according to the following scheme, in order to isolate the nonlinearities in the right-hand-side:

$$(6.54)\quad \begin{cases} (s^{n+1} - s^n)/k - \Delta s^{n+1} + \lambda s^{n+1} = \lambda[s_o - R(s^n,a^n)] \\ \\ (a^{n+1} - a^n)/k - \beta\Delta a^{n+1} + \lambda\alpha a^{n+1} = \lambda[\alpha a_o - R(s^n,a^n)] \end{cases}$$

k being a time increment, and s^n and a^n being approximations of s and a at time nk.

Then we discretize (6.54) with respect to space, by the finite difference method in the 1-dimensional case and by the finite element method in the 2-dimensional case.

The interest of the scheme (6.54) is that both equations are linear equations of the form:

$$(6.55)\quad \begin{cases} (-\mu\Delta + r)u = f \\ \\ \text{with no-flux boundary conditions} \end{cases}$$

where μ and r are strictly positive constants and the linear operator $-\mu\Delta + r$ is the same at each time level. Thus for every value of λ we only have to perform the triangularization of the matrix coming from the discretization once.

We calculate s^n and a^n for $n = 1,2,\ldots$ until we estimate that a steady state has been reached.

6.5.1 - <u>One-dimensional case</u>: A finite difference approximation of:

$$(6.56)\quad \begin{cases} -\mu u''(x) + ru(x) = f(x), & 0 < x < 1, \\ u'(0) = u'(1) = 0 \end{cases}$$

is given by:

$$(6.57)\quad \begin{cases} -\mu(u_{i+1} + u_{i-1} - 2u_i)/h^2 + ru_i = f_i & 0 \leq i \leq N \\ u_{-1} = u_1, \quad u_{N+1} = u_{N-1} \end{cases}$$

where: $h = 1/N$ is a space increment and u_i is an approximation of $u(ih)$. We have indicated in Chapter 1 how to solve such a linear system. This algorithm (Gauss elimination for a tridiagonal matrix) involves coefficients E_i and F_i $(0 \leq i \leq N)$. For every value of λ we have to perform the calculation of the coefficients E_i only once.

<u>Results</u> are shown in Figures 6.7, 6.8, 6.9, and 6.10.

<u>Figure 6.7</u> (communicated by M.C. Duban) schematically shows the non uniform steady states (N.U.S.S.) as λ monotonically increases (top), then decreases (bottom). The abscissa is λ and the ordinate, not indicated, is some quantity related to the steady state, $s(1/2)$ or $s'(1)$ for example.

In the figure, heavy lines are used for stable steady states obtained by a continuous increase in λ (top) and a continuous decrease of λ (bottom). Dashed lines are used for unstable solutions. Recall that in the interval (c_n, d_n), $(\tilde{s}, \tilde{a})$ is unstable, at least because $\sigma_n^- < 0$. Thus the uniform steady-state (U.S.S.) is stable for $0 < \lambda < c_1$ and $d_1 < \lambda < c_2$, and unstable elsewhere. Continuous increases in λ produce a family of solutions which are generally smoothly related one to the next, although there are some discontinuous jumps. The numbers $n = 1, 2, \ldots, 10$ indicate to which interval (c_n, d_n) the branch of solutions "belongs", and also the number of nodes in the concentration profiles. Since s and a profiles look the same, we have represented the s profiles only, for $n = 1, 2, 3$, and 8. Starting from $\lambda = 0$, $s = \tilde{s}$, $a = \tilde{a}$, the U.S.S. is stable until

$\lambda = c_1$. Then a polar pattern arises, corresponding to $n = 1$, one node, and with the concentrations of s above $\tilde{s}$ in one half of the spatial domain, and below $\tilde{s}$ in the other half. It is this kind of pattern which is sometimes invoked to explain head-tail differentiation. A rather uncommon feature in bifurcation diagrams is that this pattern vanishes as λ crosses d_1, and the system comes back to the trivial steady-state, which is stable again for $d_1 < \lambda < c_2$. For the particular parameter values adopted here the intervals (c_n, d_n) overlap for $n \geq 2$. But it is possible to choose these values so that the intervals (c_n, d_n) overlap only for $n \geq n_o > 2$. As an example for:

(6.58) $\alpha = 1.45$, $k = 0.1$, $s_o = 99.2$, $a_o = 76.4$, $\rho = 12.2$, and $\beta = 5$

we have:

(6.59) $z' = 0.45$, $z'' = 0.74$, and $n_o = 4$.

Coming back to Figure 6.7, for $\lambda > c_2$ the U.S.S. is no more stable, and we observe a sequence of more and more complex stable patterns with 2, then 3, ..., 8, ... nodes. At evidence the transition from n to n+1 nodes is not continuous and there are jumps from each kind of pattern to the following.

Now if λ is allowed to decrease, the system passes through stable steady states which are not necessarily the same as in the ascending phase (Figure 6.7, bottom). There is an hysteresis phenomenon, due to the existence of multiple stable steady states.

Figure 6.8 shows the profile of s for $n = 5$, i.e. with n nodes. Note the large amplitude of the deviation from $\tilde{s}$.

Figures 6.9 and 6.10 (communicated by M.C. Duban) show the s - patterns during the ascending phase for $\lambda = 20, 79, 120, 250$, and 400. Remark that, although we are far from the bifurcation point and the deviations from $\tilde{s}$ are large, the profiles still resemble cosine curves $s = \tilde{s} + A \cos n\pi x$ ($n = 1, 2, 3, 4$, and 5).

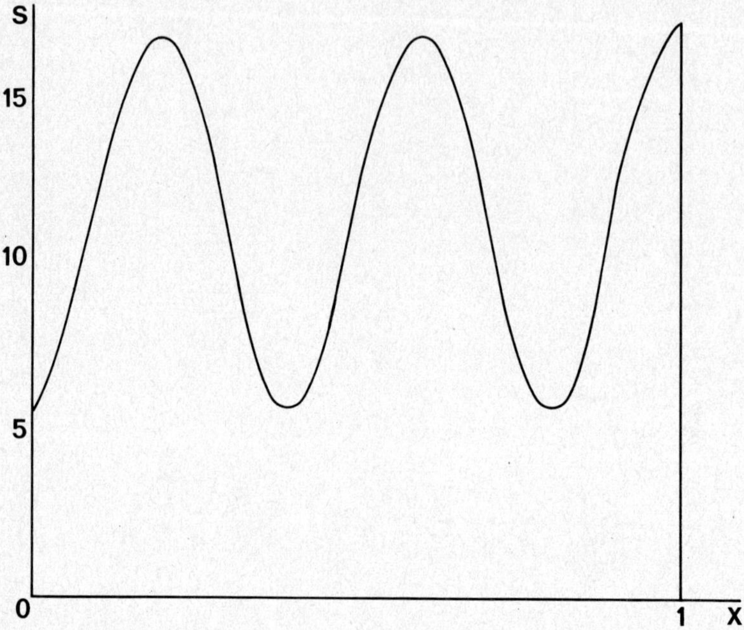

Fig. 6.8

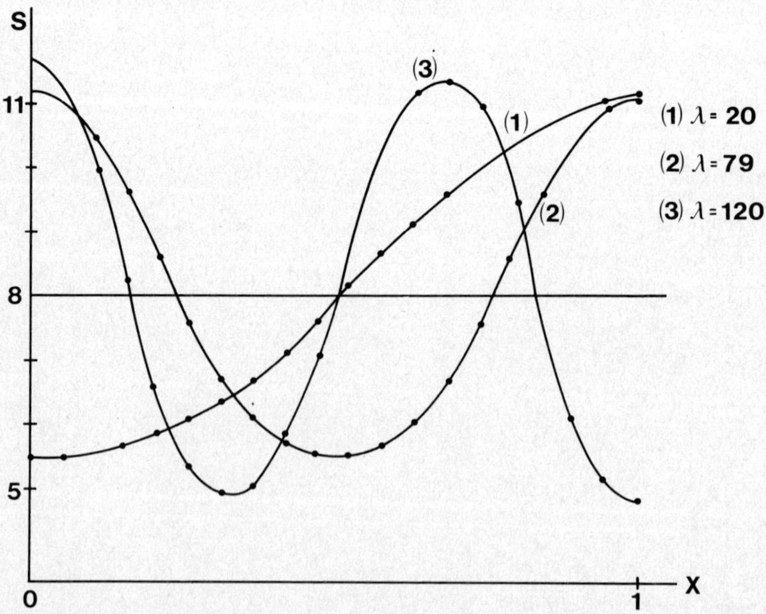

Fig. 6.9

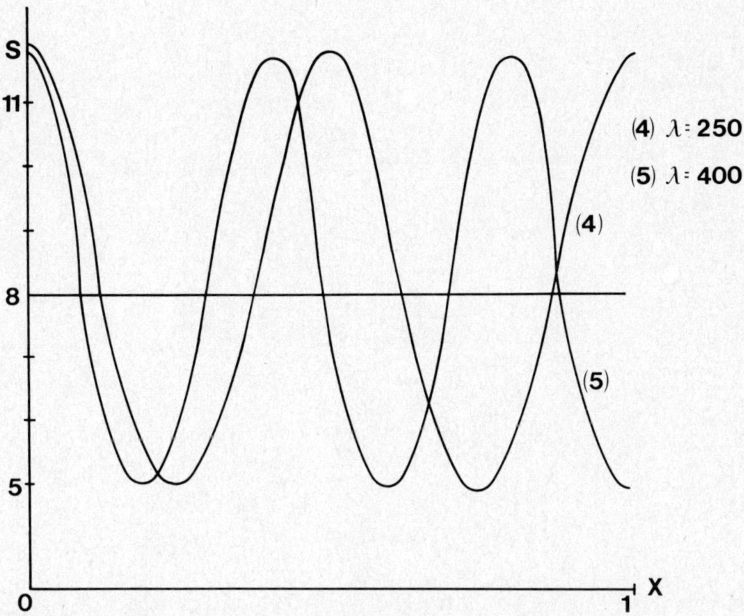

Fig. 6.10

6.5.2 - Two-dimensional case: To solve (6.55) on a planar disk Ω we discretize with respect to space by the finite element method. Let us recall briefly how the finite element method works in this case. We subdivide the spatial domain Ω into M elements Ω_m ($1 \leq m \leq M$) on each of which any function $u(x,y)$ is approximated by a polynomial expression. Figures 6.11-6.15 show such a subdivision in quadrilateral elements of spatial domains having the shape respectively of a wing disk, an eye-antenna disk, a leg disk, a genital disk, and an ellipse. The elements are 8 nodes quadrilaterals, on which piecewise biquadratic expressions are determined by their values u_i at nodes i ($1 \leq i \leq 8$) and are of the form:

$$u(x,y) = \sum_{i=1}^{8} u_i w_i(x,y)$$

where $w_i(x,y)$ is the i-th "basis function": it equals 1 at node i, 0 at other nodes, and is different from zero only in the interior of the elements, Ω_m, of which node i is a member.

Depending on the complexity of the domain, from 25 to 35 elements suffice to describe it, with between 100 and 150 total degrees of freedom in all.

Without entering into details, for which we refer to Bathe and Wilson [67], and Strang and Fix [68], let us say that an equation like (6.55) is approximated by :

$$(\mu K + rM)U = MF$$

where K is the "stiffness" matrix,

$$K_{ij} = \iint_{\Omega} (\nabla w_j)^T (\nabla w_i)\,dxdy \,, \tag{6.60}$$

M is the "mass" matrix,

$$M_{ij} = \iint_{\Omega} w_i w_j\,dxdy \,, \tag{6.61}$$

$U = (u_1, u_2, \ldots, u_N)^T$ is the vector of (unknown) degrees of freedom and $F = (f_1, f_2, \ldots, f_N)^T$ is the vector of (known) nodal values of f.

Figure 6.11 shows the discretization which was employed for the study of the wing imaginal disk. 33 elements with a total of 122 nodal points were used. For those who are familiar with finite difference methods for partial differential equations, this discretization seems very coarse. However, the smoothness of the basis functions in the finite element method is such that a coarse discretization is satisfactory as long as the maximum spatial frequency in the exact solution is low enough.

Coming back to (6.54) we have to solve, at each step of time, in order to calculate s^{n+1} and a^{n+1} from s^n and a^n, two linear systems:

$$\begin{cases} K_1 s^{n+1} = MF_1^n \\ K_2 a^{n+1} = MF_2^n \end{cases}$$

where:

$$K_1 = K + (\lambda + 1/k)M$$

$$K_2 = \beta K + (\alpha\lambda + 1/k)M$$

while:

$$(F_1^n)_i = \lambda[s_o - R(s_i^n, a_i^n)]$$

$$(F_2^n)_i = \lambda[\alpha a_o - R(s_i^n, a_i^n)] \quad 1 \leq i \leq N.$$

For every value of λ we only have to assemble the two matrices K and M and to perform the triangularizations of K_1 and K_2 once and for all time steps. For every value of λ we take as starting values s^o, a^o, the steady state obtained for the preceding value of λ, slightly perturbed by random noise, and calculate s^n and a^n for n = 1,2,... until we estimate that a steady state has been reached again.

Results - We have already shown in Figure 5.1 a selection of the solutions obtained in this manner on an ellipse. Since both s and a show the same concentration pattern, only s needs to be represented. Here as well as in the subsequent figures, darker plotting represent larger values. These patterns, obtained by G. Joly [46], are the stable steady states, as λ increases, for λ = 0.3, 0.6, 0.7, 2., 3.5, and 5, (top to bottom, right to left).

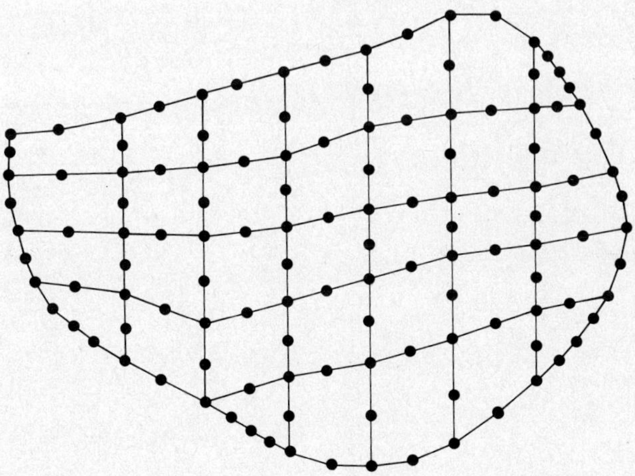

Fig. 6.11

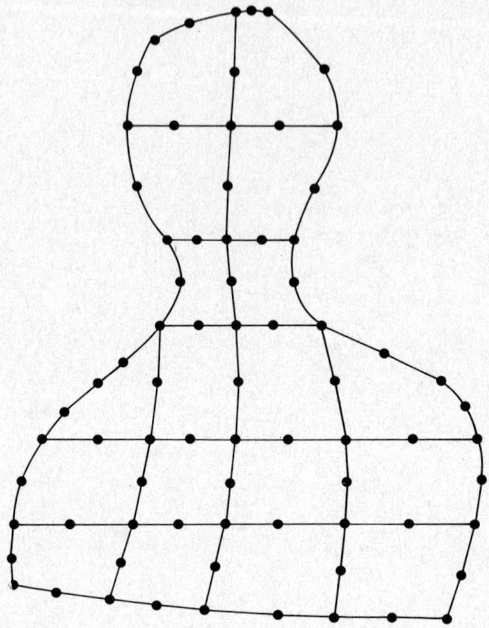

Fig. 6.12

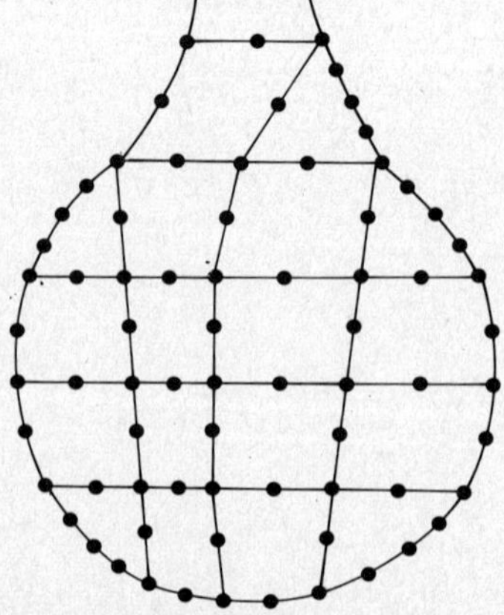

Fig. 6.13

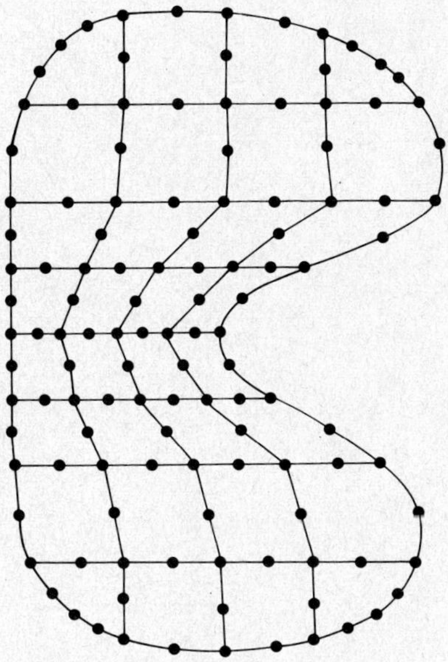

Fig. 6.14

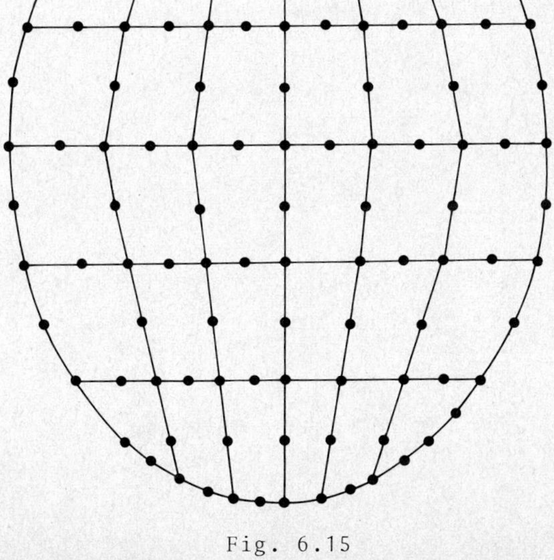

Fig. 6.15

Fig. 6.17

Fig. 6.18

Fig. 6.19

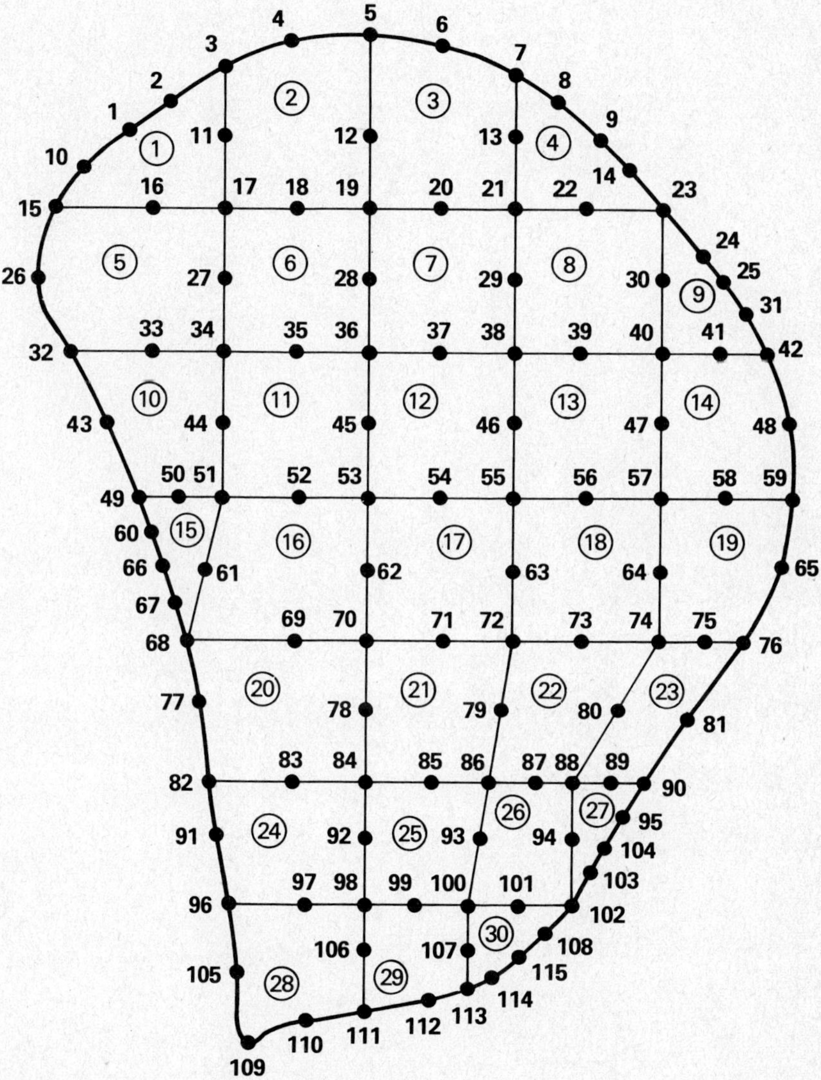
1
2
3
4
5
6
7
8
9
10
11
12
13
14
15
16
17
18
19
20
21
22
23
24
25
26
27
28
29
30
31
32
33
34
35
36
37
38
39
40
41
42
43
44
45
46
47
48
49
50
51
52
53
54
55
56
57
58
59
60
61
62
63
64
65
66
67
68
69
70
71
72
73
74
75
76
77
78
79
80
81
82
83
84
85
86
87
88
89
90
91
92
93
94
95
96
97
98
99
100
101
102
103
104
105
106
107
108
109
110
111
112
113
114
115
(1)
(2)
(3)
(4)
(5)
(6)
(7)
(8)
(9)
(10)
(11)
(12)
(13)
(14)
(15)
(16)
(17)
(18)
(19)
(20)
(21)
(22)
(23)
(24)
(25)
(26)
(27)
(28)
(29)
(30)

Fig. 6.20

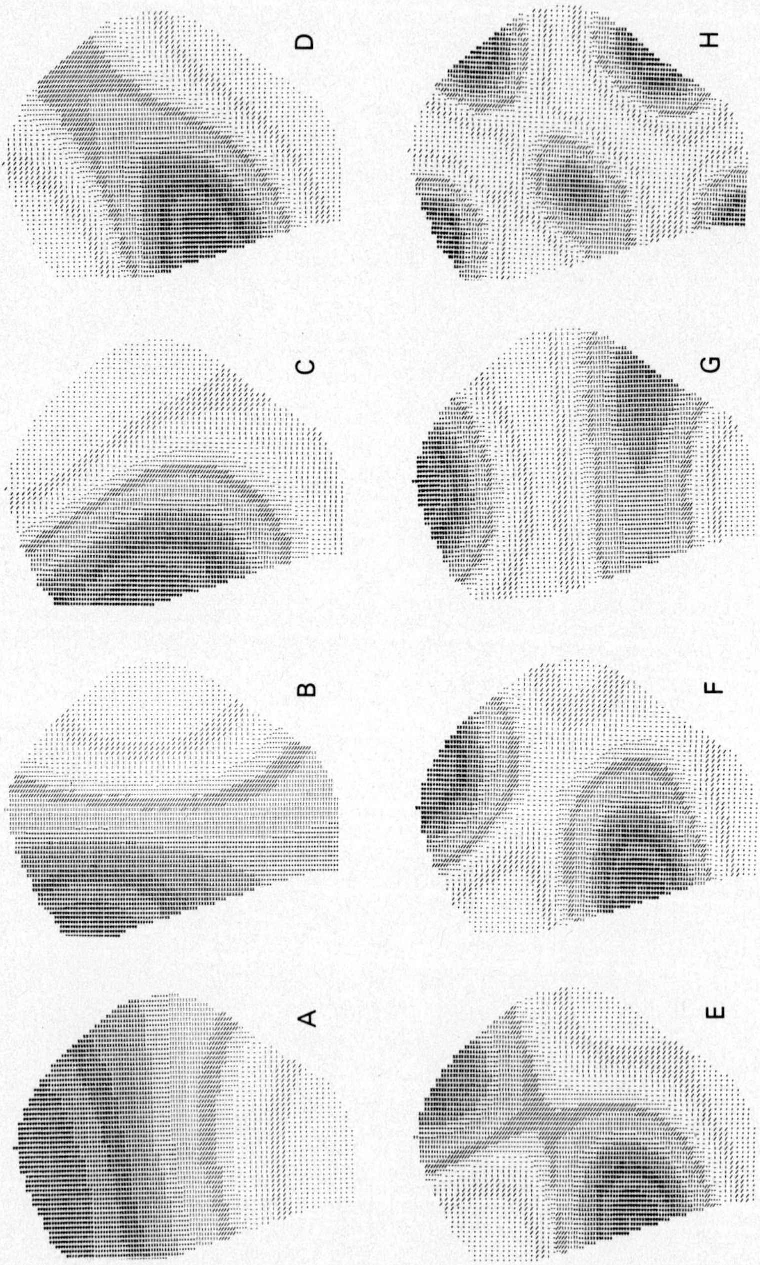

Fig. 6.21

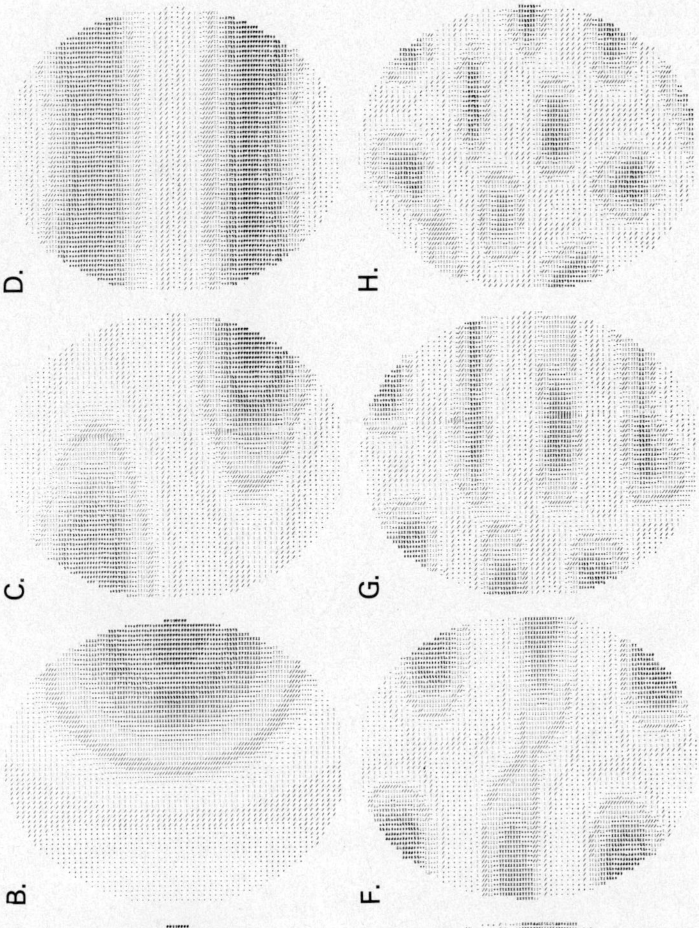

Fig. 6.22

6.5.3 - Case of a surface: The s-a system more generally can represent diffusion and enzyme reaction on a surface S separated from an external medium by a boundary layer. The Laplace-Beltrami operator Δ_B is then appropriate to represent diffusion on the surface, and the s-a system equations are:

$$(6.62)\quad \begin{cases} s_t - \Delta_B s + \lambda[R(s,a) - (s_o - s)] = 0 \\ \\ a_t - \beta\Delta_B a + \lambda[R(s,a) - \alpha(a_o - a)] = 0 \end{cases}$$

with no boundary conditions if the surface is closed. Let us recall the definition of the Laplace-Beltrami operator: locally the surface S (think of a cucumber surface!) can be mapped into the interior of a square: $-1 < r,s < +1$ (Figure 6.23) and x,y,z coordinates on S can be expressed as:

$$(6.63)\quad x = x(u_1,u_2), \quad y = y(u_1,u_2), \quad z = z(u_1,u_2).$$

Define:

$$(6.64)\quad dl^2 = dx^2 + dy^2 + dz^2 = a_{11}du_1^2 + 2a_{12}du_1du_2 + a_{22}du_2^2,$$

$$(6.65)\quad a = a_{11}a_{22} - a_{12}^2$$

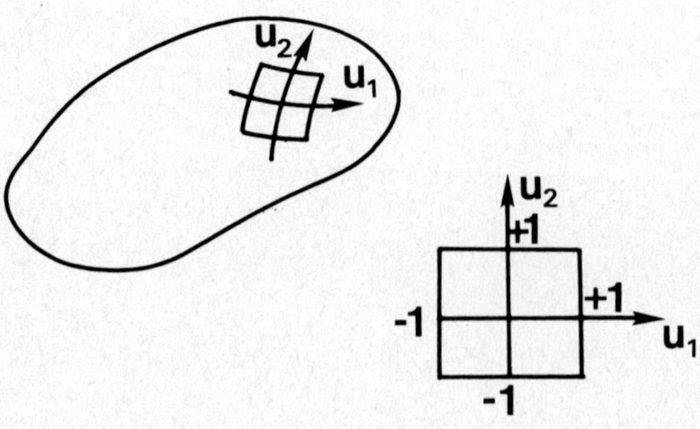

Fig. 6.23

$$(6.66)\quad a^{11} = \frac{a_{22}}{a}, \quad a^{22} = \frac{a_{11}}{a}, \quad a^{12} = a^{21} = -\frac{a_{12}}{a}.$$

Then:

(6.67) $dS = a^{1/2}\, du_1 du_2$

(6.68) $\Delta_B \phi = \frac{1}{a^{1/2}} \sum_{\alpha,\beta} \frac{\partial}{\partial u_\beta} (a^{1/2}\, a^{\alpha\beta} \frac{\partial \phi}{\partial u_\alpha})$

(6.69) $\int_S (-\Delta_B \phi)\, \psi \, dS = \int_S a^{\alpha\beta} \frac{\partial \phi}{\partial u_\alpha} \frac{\partial \psi}{\partial u_\beta}\, dS.$

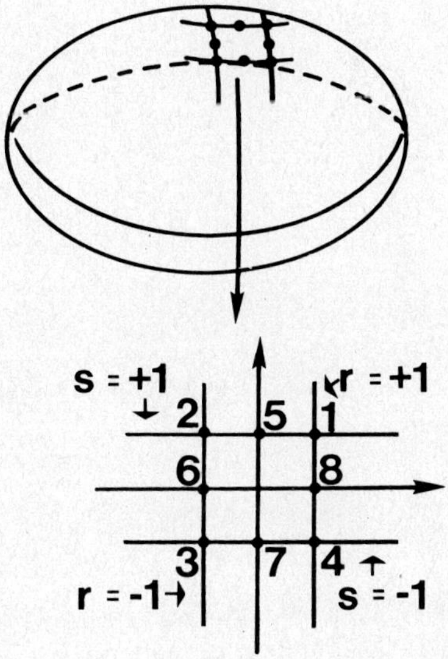

Fig. 6.24

It is no more difficult to solve numerically the s-a system on a cucumber surface than on a planar surface by using the finite element method. However we have in mind the surface of an egg rather than the surface of a cucumber.

Let us give some details on the calculation of the "stiffness" and "mass" matrices K and M in the case of a surface (a planar domain being a particular case). The surface is divided into curved 8-nodes quadrilaterals (see Figure 6.24). The nodes are defined by their coordinates (x_i, y_i, z_i for node number i of the quadrilateral, $1 \leq i \leq 8$). We call $H_i(r,s)$ the "shape functions":

$$(6.70)\quad \begin{cases} H_1(r,s) = \frac{1}{4}(1+r)(1+s) - \frac{1}{2}H_5(r,s) - \frac{1}{2}H_8(r,s) \\ H_2(r,s) = \frac{1}{4}(1-r)(1+s) - \frac{1}{2}H_6(r,s) - \frac{1}{2}H_5(r,s) \\ H_3(r,s) = \frac{1}{4}(1-r)(1-s) - \frac{1}{2}H_7(r,s) - \frac{1}{2}H_6(r,s) \\ H_4(r,s) = \frac{1}{4}(1+r)(1-s) - \frac{1}{2}H_8(r,s) - \frac{1}{2}H_7(r,s) \\ H_5(r,s) = \frac{1}{2}(1-r^2)(1+s) \\ H_6(r,s) = \frac{1}{2}(1-s^2)(1-r) \\ H_7(r,s) = \frac{1}{2}(1-r^2)(1-s) \\ H_8(r,s) = \frac{1}{2}(1-s^2)(1+r) \end{cases}$$

The part of the surface delineated by the quadrilateral is approximated by the surface $S^{(m)}$ whose parametric representation is:

$$(6.71)\quad \begin{cases} x = \sum_{i=1}^{8} x_i H_i(r,s), y = \sum_{i=1}^{8} y_i H_i(r,s), \quad z = \sum_{i=1}^{8} z_i H_i(r,s) \\ -1 \leq r,s \leq +1. \end{cases}$$

The elementary mass and stiffness matrices are:

$$(6.72)\quad K^{(m)} = \int_{S^{(m)}} B^{(m)^T} D^{(m)} B^{(m)} dS^{(m)} \quad \text{and} \quad M^{(m)} = \int_{S^{(m)}} H^{(m)^T} H^{(m)} dS^{(m}$$

where $dS^{(m)}$, $B^{(m)}$, $D^{(m)}$, and $H^{(m)}$ are defined below.

To each pair $(r,s) \in [-1,+1] \times [-1,+1]$ corresponds a point (x,y,z) on $S^{(m)}$ by equations (6.71) and we call:

$$(6.73)\quad H_i^{(m)}(x,y,z) = H_i(r,s) \qquad 1 \leq i \leq 8.$$

Thus $H_i^{(m)}$ is a function of the point (x,y,z) on the surface $S^{(m)}$. $H^{(m)}$ denotes the vector:

$$H^{(m)} = [H_1^{(m)}, H_2^{(m)}, \ldots, H_8^{(m)}].$$

Similarly we define:

$$(6.74) \quad B^{(m)} = \begin{bmatrix} \dfrac{\partial H_1}{\partial r} & \cdots & \dfrac{\partial H_8}{\partial r} \\ \\ \dfrac{\partial H_1}{\partial s} & \cdots & \dfrac{\partial H_8}{\partial s} \end{bmatrix}$$

$H^{(m)}$ and $B^{(m)}$ are 1 x 8 and 2 x 8 matrices defined on $S^{(m)}$. Now, following equations (6.64)-(6.67) we define:

$$dl^2 = dx^2 + dy^2 + dz^2 = \left(\sum_{i=1}^{8} x_i \left(\frac{\partial H_i}{\partial r} dr + \frac{\partial H_i}{\partial s} ds\right)\right)^2$$

$$+ \left(\sum_i y_i \left(\frac{\partial H_i}{\partial r} dr + \frac{\partial H_i}{\partial s} ds\right)\right)^2 + \left(\sum_i z_i \left(\frac{\partial H_i}{\partial r} dr + \frac{\partial H_i}{\partial s} ds\right)\right)^2$$

$$= a_{11} dr^2 + 2a_{12} drds + a_{22} ds^2$$

$$(6.75) \quad \begin{cases} a_{11} = \vec{V}_1^{\,2}, \quad a_{12} = \vec{V}_1 \vec{V}_2, \quad a_{22} = \vec{V}_2^{\,2} \\ \vec{V}_1 = \left[\sum_{i=1}^{8} x_i \dfrac{\partial H_i}{\partial r}, \ \sum y_i \dfrac{\partial H_i}{\partial r}, \ \sum z_i \dfrac{\partial H_i}{\partial r}\right]^T \\ \vec{V}_2 = \left[\sum x_i \dfrac{\partial H_i}{\partial s}, \ \sum y_i \dfrac{\partial H_i}{\partial s}, \ \sum z_i \dfrac{\partial H_i}{\partial s}\right]^T. \end{cases}$$

Then we define:

$$(6.76) \quad \begin{cases} a = a_{11} a_{22} - a_{12}^2 \\ a^{11} = a_{22}/a, \quad a^{22} = a_{11}/a, \quad a^{12} = -a_{12}/a \end{cases}$$

(6.77) $dS^{(m)} = a^{1/2} dr ds$,

(6.78) $$D^{(m)} = \begin{bmatrix} a^{11} & a^{12} \\ a^{12} & a^{22} \end{bmatrix},$$

(6.79) $$K^{(m)} = \int_{-1 \le r, s \le +1} B^T D^{(m)} B \, a^{1/2} \, dr ds,$$

(6.80) $$M^{(m)} = \int_{-1 \le r, s \le +1} H^T H \, a^{1/2} \, dr ds$$

Using this method and the program SSPACE (subspace iteration method) of Bathe and Wilson [67], the first eigenvalues and eigenfunctions of the Laplace Beltrami operator $-\Delta_B$ on the surface of an elongated ellipsoid with the form of an egg can be obtained (Bunow et.al. [69]).

The first one corresponds to the onset of polarity (distinction head-tail), the second one to the dorsal-ventral separation and the third one to the left-right differentiation (Figure 6.25). Figure 6.26, provided by B. Bunow, shows the sequence of eigenfunctions on the surface of a Drosophila egg at the cellular blastoderm stage. The left and right surfaces of the egg have been projected onto the sagittal plane. The first eigenfunction is shown at the bottom of the first two columns. From bottom to top of these two columns we have the first four eigenfunctions each column corresponding to a side of the egg. Across the 3rd and 4th columns, from bottom to top, we have the next five eigenfunctions.

It results from our previous analysis of bifurcation from the uniform state in the s-a system that it is possible to find patterns similar to these first eigenfunctions when solving the nonlinear equations (6.62) instead of the eigenvalue problem.

Surfaces where the diffusion coefficients are varying (because of a layer of cells of variable size for instance) model more closely the shape of the real imaginal discs (Figure 6.27). It is thus possible to solve the s-a system equations on such surfaces and study the sequence of stable patterns.

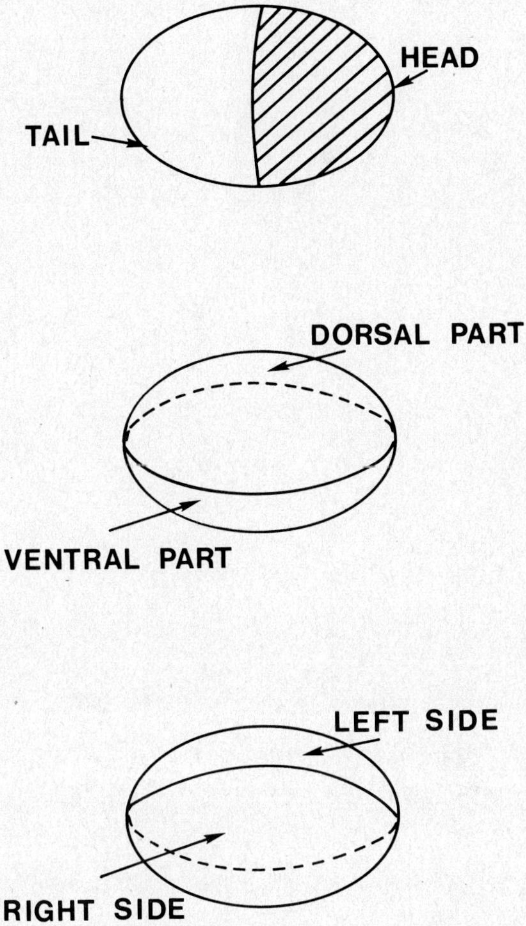

Fig. 6.25

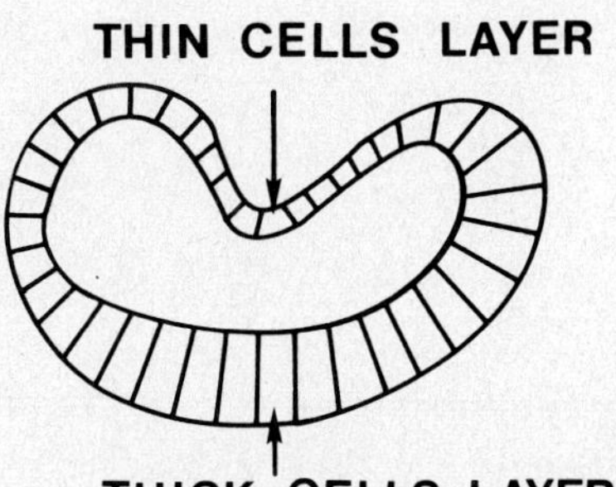

Fig. 6.27

Fig. 6.26

If the diffusion coefficients are varying on a surface, we only have to multiply the entries a^{ij} of the matrix $D^{(m)}$ in (6.78) by an appropriate coefficient.

This does not complicate the calculation of the stiffness matrix K.

Thus the finite element method is a very efficient tool which allows us to solve the s-a system on any kind of surface S, planar or not, with any field of diffusion coefficients. Moreover it enables us to calculate the first few eigenvalues and eigenfunctions:

$$(6.81) \quad -\nabla(D\nabla w_i) = \mu_i w_i, \qquad 0 \leq i \leq M$$

on S with eventually no-flux boundary conditions if S is not closed. Here $D = D(x)$ may be a variable coefficient.

This knowledge of the first few eigenvalues and eigenfunctions is the basis of a method which saves computer time, and that we are going to describe now, namely Galerkin's method.

6.5.4 - <u>Galerkin's method</u>: We explain Galerkin's method for the evolution problem (6.1) on the wing disk of Figure 6.11. Here we pick out the first M eigenvalues and eigenfunctions defined by (6.81). Of course we work with time and space discretizations, but we prefer explaining the method on the original problem (6.1).

We define approximations s_M and a_M of s and a by:

$$(6.82) \quad s_M(t) \text{ and } a_M(t) \in \text{span } \{w_o, w_1, \dots, w_M\} .$$

More precisely we restrict ourselves to the problem of finding the $2(M+1)$ functions $\xi_i(t)$ and $\eta_i(t)$, $0 \leq i \leq M$, such that:

$$(6.83) \quad \begin{cases} s_M(t) = \sum_{i=0}^{M} \xi_i(t) w_i, \quad a_M(t) = \sum_{i=0}^{M} \eta_i(t) w_i \\ \int_\Omega \{s'_M - \Delta s_M + \lambda[R(s_M, a_M) - (s_o - s_M)]\} w_i \, dx = 0 \\ \qquad\qquad\qquad\qquad\qquad\qquad\qquad\qquad\qquad 0 \leq i \leq M \\ \int_\Omega \{a'_M - \beta \Delta a_M + \lambda[R(s_M, a_M) - \alpha(a_o - a_M)]\} w_i \, dx = 0 \end{cases}$$

where of course the eigenfunctions w_i are functions of the spatial variable x,

$$s'_M = \sum_{i=0}^{M} \xi'_i(t)w_i, \qquad -\Delta s_M = \sum_{i=0}^{M} \mu_i \xi_i(t) w_i,$$

$$a'_M = \sum_{i=0}^{M} \eta'_i(t)w_i \quad \text{and} \quad -\Delta \ _M = \sum_{i=0}^{M} \mu_i \eta_i(t) w_i.$$

Equations (6.83) are in fact a system of 2(M+1) ordinary differential equations:

$$(6.84)\quad \begin{cases} \xi'_i + (\mu_i + \lambda)\xi_i - \lambda s_o \delta_{oi} + T_i = 0 & \\ \eta'_i + (\beta\mu_i + \lambda\alpha)\eta_i - \lambda\alpha a_o \delta_{oi} + T_i = 0 & 0 \leq i \leq M \\ T_i = \lambda \int_\Omega R\left(\sum_{j=0}^{M} \xi_j w_j, \sum_{j=0}^{M} \eta_j w_j \right) w_i \, dx. & \end{cases}$$

Here we have used the normalization condition:

$$(6.85)\qquad \int_\Omega w_i w_j \, dx = \delta_{ij}.$$

This method has been employed by G. Joly [71].

Taking M = 5, she finds, for λ increasing continuously from 0 to 200, the same stable patterns with 2(M+1) = 12 degrees of freedom than with 2 x 122 = 244 degrees of freedom in the previous method (2 degrees of freedom s_i and a_i for each of the 122 nodal points). Of course a preliminary calculation of the eigenpairs (μ_i, w_i) is necessary, but they can be calculated once and for all for a lot of applications involving the solution of the evolution equations as previously described or the continuation of solutions as in the preceding chapter. For details we refer to [71].

6.6 Mathematical Technical Details

We delayed until now the justification of some important mathematical properties. In the present Section we essentially investigate the properties of the operator:

$$(6.86)\quad f_u(\lambda,u_o) = \begin{bmatrix} -\Delta + \lambda(\tilde{R}_s+1) & \lambda R_a \\ \lambda\tilde{R}_s & -\beta\Delta + \lambda(\tilde{R}_a+\alpha) \end{bmatrix}$$

We first show that its spectrum is only composed of eigenvalues with finite multiplicities. Then we give a linarized stability theorem for the proof of which we refer to [72], and apply it to the stability analysis of the trivial state $(\tilde{s},\tilde{a})$. Finally we prove that f is a twice continously differentiable mapping, the range of which is closed and of codimension 1.

6.6.1 - Spectrum of $f_u(\lambda,u_o)$: We generalize here some of the methods employed by Meurant and Saut [34] in the study of bifurcation and stability in the Bruxellator. We are within the framework of Section 6.3, with X and Y defined by (6.21)-(6.23) and $f : R^+ \times X \to Y$ defined by (6.20). The spatial domain Ω is supposed to be an open bounded set in R^n (n = 1, 2, or 3). We denote $(\cdot,\cdot)$ the scalar product in Y (or in $L^2(\Omega)$, no ambiguity being possible), and $|\cdot|$ the corresponding norm. Recall u_o is the trivial state $(\tilde{s},\tilde{a})$. We define the operators L_o and L_1 as:

$$(6.87)\quad L_o = \begin{bmatrix} -\Delta & 0 \\ 0 & -\beta\Delta \end{bmatrix} \quad\text{and}\quad L_1 = \begin{bmatrix} \tilde{R}_s+1 & \tilde{R}_a \\ \tilde{R}_s & \tilde{R}_a+\alpha \end{bmatrix}$$

and

$$(6.88)\quad L_\lambda = L_o + \lambda L_1 = f_u(\lambda,u_o).$$

Definition 6.1 - Let T be a linear operator from $D(T) = X \subset Y$ into Y. The "numerical range" $\Theta(T)$ of T is the set of those $z \in \mathbb{C}$ such that:

$$z = (TU,U), \quad U \in X, \quad |U| = 1.$$

Example 6.1 - Let $U = (u,v) \in X$, X defined in (6.22) and (6.23), and $T = L_\lambda$. We have:

$$(L_\lambda U,U) = [u]^2 + \beta[v]^2 + \lambda[(\tilde{R}_s+1)|u|^2 + (\tilde{R}_a+\alpha)|v|^2 + R_a(v,\bar{u}) + R_s(u,\bar{v})]$$

where the functions u and v may be complex-valued and,

$$[u]^2 = \sum_{i=1}^{n} \int_\Omega \frac{\partial u}{\partial x_i} \frac{\partial \bar{u}}{\partial x_i} dx.$$

Definition 6.2 - A linear operator T is said to be "sectorial", with vertex γ and half-angle θ if the numerical range $\Theta(T)$ is included in the complex sector:

(6.89) $$\Sigma: \mathrm{Re}\, z \geq \gamma, \quad |\mathrm{Im}\, z| \leq \mathrm{tg}\, \theta\, (\mathrm{Re}\, z - \gamma).$$

Example 6.2 - Since $\tilde{R}_s < 0$ and $\tilde{R}_a > 0$, we have:

$$\mathrm{Re}(L_\lambda U,U) = [u]^2 + \beta[v]^2 + \lambda(\tilde{R}_s+1)|u|^2 + \lambda(\tilde{R}_a+\alpha)|v|^2$$

$$+ \lambda\tilde{R}_a \mathrm{Re}(v,\bar{u}) + \lambda\tilde{R}_s \mathrm{Re}(u,\bar{v}) \geq \lambda\tilde{R}_s|u|^2 - \lambda(\tilde{R}_a-\tilde{R}_s)|u||v|,$$

and

$$|\mathrm{Im}(L_\lambda U,U)| = \lambda|\mathrm{Im}(\tilde{R}_a(v,\bar{u}) + \tilde{R}_s(u,\bar{v}))| \leq \lambda(\tilde{R}_a-\tilde{R}_s)|u||v|.$$

Thus:

$$\mathrm{Re}(L_\lambda U,U) + \lambda(\tilde{R}_a-2\tilde{R}_s)(|u|^2+|v|^2) \geq \lambda(\tilde{R}_a-\tilde{R}_s)|u||v| \geq |\mathrm{Im}(L_\lambda U,U)|$$

and we finally have:

$$\mathrm{Re}((L_\lambda-\gamma_\lambda I)U,U) \geq |\mathrm{Im}(L_\lambda U,U)|$$

where: $\gamma_\lambda = 2\tilde{R}_s - \tilde{R}_a$, whence the following Lemma:

Lemma 6.1 - L_λ is a sectorial operator with vertex $\gamma_\lambda = 2\tilde{R}_s - \tilde{R}_a$ and half-angle $\pi/4$.

Lemma 6.2 - If $\delta \geq \gamma_\lambda$, then $(L_\lambda + \delta I)^{-1}$ is compact.

Proof - Since $L_o^{-1} : Y \to Y$ is compact and $L_1 : Y \to Y$ is bounded, $L_1 L_o^{-1}$ is compact and so is $A = \lambda L_1 L_o^{-1} + \delta L_o^{-1} : Y \to Y$.

Since $\mathrm{Re}((L_\lambda - \gamma_\lambda I)U,U) \geq 0$ and $\delta + \gamma_\lambda > 0$ we have :

$$\mathrm{Re}((L_\lambda + \delta I)U,U) \geq (\delta+\gamma_\lambda)|U|^2 \text{ and } L_\lambda + \delta I \text{ is injective.}$$

In other words $L_o U + \lambda L_1 U + \delta U = 0 \Rightarrow U = 0$, or $AU = (\lambda L_1 L_o^{-1} + \delta L_o^{-1})U = -U$ admits the single solution $U = 0$.

Since -1 is not an eigenvalue of the compact operator A, it belongs to its resolvent set. Thus $(I+A)^{-1}$ is bounded. It follows that:

$$(L_\lambda + \delta I)^{-1} = [(I+A)L_o]^{-1} = L_o^{-1}(I+A)^{-1} : Y \to Y$$

is compact as the product of a bounded operator and a compact operator.

Lemma 6.3 - The resolvent of L_λ is compact.

Proof - It is well known [73,74] that if T is a closed operator in a Banach space Y such that the resolvent $(\zeta I - T)^{-1}$ exists and is compact for some ζ, then the spectrum of T entirely consists of isolated eigenvalues with finite multiplicities and $(\zeta I - T)^{-1}$ is compact for all $\zeta \in \rho(T)$ (resolvent set of T). Here $T = L_\lambda$ is closed, as the sum of a closed operator in Y, L_o, and a bounded operator in Y, λL_1. Moreover $(-\delta I - L_\lambda)^{-1}$ exists and is compact. This completes the proof of the Lemma, and proves also our final result:

Proposition 6.1 - The spectrum of L_λ is discrete and only composed of eigenvalues having finite multiplicity.

6.6.2 - The principle of linearized stability: Let u_o be an equilibrium point of a dynamical system :

$$\frac{du}{dt} + f(u) = 0.$$

Suppose:

$$\mathrm{Re}\,\sigma(f_u(u_o)) > 0,$$

which means that the real part of the elements of the spectrum of $f_u(u_o)$, $\sigma(f_u(u_o))$, is strictly positive. Can we assert that u_o is stable? The answer is yes, provided $f_u(u_o)$ is an m-sectorial operator. Before stating more precisely this property, we must define what we mean by m-sectorial operators, and also the norm $||\cdot||_\alpha$ under which

$$\|u(t) - u_o\|_\alpha \to 0 \text{ as } t \to \infty, \text{ provided } \|u(0) - u_o\|_\alpha < \varepsilon.$$

Definition 6.3 - A linear operator T in a Banach space Y is said to be m-accretive if, for $\mathrm{Re}\sigma > 0$,

$$(T + \sigma I)^{-1} \in L(Y) \text{ and } \|(T + \sigma I)^{-1}\| \leq 1/\mathrm{Re}\sigma.$$

Here $L(Y)$ is the space of bounded linear operators in Y and $\|\cdot\|$ is the norm in $L(Y)$.

Definition 6.4 - T is said to be quasi-m-accretive if, for some real c, T + cI is m-accretive.

Definition 6.5 - We call T an m-sectorial operator if it is sectorial and quasi-m-accretive.

Example 6.3 - Define:

$$a(u,v) = \sum_{i=1}^{n} \int_\Omega \frac{\partial u}{\partial x_i} \frac{\partial v}{\partial x_i} \, dx + \int_\Omega uv \, dx \text{ for u and } v \in H^1(\Omega),$$

and consider those u such that there exists a function $f \in L^2(\Omega)$ such that:

$$a(u,v) = (f,v) \qquad \forall v \in H^1(\Omega).$$

Let $A : D(A) \to L^2(\Omega)$ be the linear mapping defined by Au = f:

$$(Au,v) = a(u,v) \quad \forall v \in H^1(\Omega).$$

It is well-known [75] that the set of those u such that the mapping $v \to a(u,v)$ is continuous for the topology induced on $H^1(\Omega)$ by $L^2(\Omega)$, is:

$$D(A) = \{u | u \in H^1(\Omega) \text{ and } \Delta u \in L^2(\Omega)\},$$

and, if $\partial\Omega$ is smooth enough,

$$D(A) = \{u | u \in H^2(\Omega) \text{ and } \frac{\partial u}{\partial \nu} = 0\} = V.$$

It is also well- known that, if $u \in D(A)$, then $Au = -\Delta u + u$, and, endowed with the graph norm, D(A) is a Banach space:

$$||u||_{D(A)} = |Au|_{L^2(\Omega)} = |-\Delta u+u|_{L^2(\Omega)}.$$

The operator A is m-accretive. Indeed, if $Re\sigma > 0$ and $f \in L^2(\Omega)$, the equation $Au + \sigma u = f$ admits an unique solution, satisfying:

$$a(u,v) + \sigma(u,v) = (f,v) \quad \forall v \in H^1(\Omega),$$

whence, for $v = \bar{u}$,

$$[u]^2 + (1+\sigma)|u|^2 = (f,\bar{u}).$$

Taking the real part, we have:

$$[u]^2 + (1+Re\sigma)|u|^2 = Re(f,\bar{u}) \leq |f||u|$$

and in particular:

$$|u| \leq |f|/Re\sigma$$

whence:

$$||(A+\sigma)^{-1}||_{\mathcal{L}(Y)} \leq 1/Re\sigma \ , \quad Y = L^2(\Omega).$$

Example 6.4 - Consider the operator L_o defined in (6.87), with $D(L_o) = V \times V$, and consider the equation $(L_o + cI + \sigma I)U = G$, where $c \in R^+$, $Re\sigma > 0$, $U = (u,v)$ and $G = (g,h) \in Y$. This equation splits into:

$$\begin{cases} -\Delta u + (c+\sigma)u = g, & \dfrac{\partial u}{\partial \nu} = 0 \\ -\beta\Delta v + (c+\sigma)v = h, & \dfrac{\partial v}{\partial \nu} = 0 \end{cases}$$

and it results from example 6.3 that $|u| \leq |g|/Re\sigma$ and $|v| \leq |h|/Re\sigma$, whence $|u|^2 + |v|^2 \leq (|g|^2+|h|^2)/(Re\sigma)^2$, that is $|U| \leq |G|/Re\sigma$, so that:

$$||(L_o+cI+\sigma I)^{-1}||_{\mathcal{L}(Y)} \leq 1/Re\sigma, \qquad Y = L^2(\Omega) \times L^2(\Omega).$$

Therefore L_o is quasi-m-accretive. Moreover it is clear that L_o is sectorial with vertex 0 and half-angle any $\theta \in (0,\pi/2)$. Thus L_o is m-sectorial.

If T is an m-sectorial operator in a Hilbert space Y, then it is possible to define Banach spaces Y^α $(\alpha \geq 0)$. For a general definition of these spaces we refer to [72,73,74]. For the particular case we are interested in, where $Y = L^2(\Omega) \times L^2(\Omega)$ and T is the operator L_o of example 6.4, it suffices to say that Y^α can be defined as follows:

Let (μ_i, w_i) be the eigenpairs of the operator A of example 6.3:

$$\begin{cases} -\Delta w_i + w_i = \mu_i w_i, \\ \dfrac{\partial w_i}{\partial \nu} = 0 \quad , \qquad\qquad i = 0,1,2,\ldots \quad . \end{cases}$$

If u and $v \in L^2(\Omega)$, they can be written:

$$u = \sum_{i=0}^{\infty} u_i w_i \quad \text{and} \quad v = \sum_{i=0}^{\infty} v_i w_i .$$

The pair $U = (u,v)$ will be said to belong to Y^α if:

$$\sum_{i=0}^{\infty} \mu_i^{2\alpha} (|u_i|^2 + |v_i|^2) < \infty.$$

Y^α is a Banach space with the norm:

$$\|U\|_\alpha = \Big(\sum_{i=0}^{\infty} \mu_i^{2\alpha}(|u_i|^2 + |v_i|^2)\Big)^{1/2}.$$

In particular $Y^0 = Y$ and $Y^1 = D(L_o) = X$. We consider now the following situation. Let A be an m-sectorial operator in a Hilbert space Y, and let $r : U \to Y$ where U is a neighborhood in Y^α (some $\alpha < 1$) of u_o. We say u_o is an equilibrium point of the dynamical system:

$$(6.90) \quad \frac{du}{dt} + Au = r(u)$$

if $u_o \in D(A)$ and $Au_o = r(u_o)$.

An equilibrium solution u_o is stable (in Y^α) if, for any $\varepsilon > 0$, there exists $\delta > 0$ such that any solution u of (6.90) with $\|u(0) - u_o\|_\alpha < \delta$

satisfies $\|u(t) - u_o\|_\alpha < \varepsilon$ for all $t > 0$. We say u_o is _asymptotically stable_ if it is stable and we can also guarantee that $u(t) \to u_o$ in Y^α as $t \to \infty$ by choosing $\|u(0) - u_o\|_\alpha$ sufficiently small. We say u_o is _unstable_ if it is not stable.

We assume f is locally Lipschitzian on U. More precisely, if $u_1 \in U$, there exists a neighborhood $U_1 \subset U$ of u_1 such that for $x \in U_1$, $v \in U_1$,

$$|r(x) - r(y)| \leq L \|x-y\|_\alpha$$

for some constant $L > 0$.

Suppose:

$$r(u_o+z) = r(u_o) + r'(u_o)z + g(z)$$

where $r'(u_o)$ is a bounded linear map from Y^α to Y, and:

(6.91) $|g(z)| = O(\|z\|^2_\alpha)$ as $\|z\|_\alpha \to 0$.

Let L be the "linearized operator" from X to Y:

$$L : w \to Aw + r'(u_o)w.$$

The following results have been obtained in a more general framework in [72], to which we refer for the proofs.

Theorem 6.2 - With the above assumptions on r, u_o, and u_1, we have the following:

(i) (_Existence and uniqueness_) (6.90) has an unique solution on $[0,+\infty[$ with initial value $u_1 \in Y^\alpha$. It is a continuous function $u : [0,\infty[\to Y$ such that $u(0) = u_1$ and on $]0,\infty[$ we have $u(t) \in D(A)$.

(ii) (_Stability_) If the spectrum of L lies in $\{\mathrm{Re} z > \beta\}$ for some $\beta > 0$, then u_o is asymptotically stable. More precisely, there exist $\delta > 0$ and $M \geq 1$ such that if $\|u_1 - u_o\|_\alpha < \delta$, then on $0 \leq t < \infty$ the solution of (6.90) satisfies:

$$\|u(t) - u_o\|_\alpha \leq 2Me^{-\beta t} \|u_1 - u_o\|_\alpha .$$

(iii) (Asymptotic behavior) Assume that $\sigma(L) \subset \{\beta\} \cup \{\mathrm{Re}\,z > \beta'\}$, where β is a simple positive eigenvalue of L and $\beta' > \beta > 0$. Then for any γ in $\beta < \gamma < \min(\beta', 2\beta)$, there exists $\delta > 0$ so that $\|u_1 - u_o\|_\alpha < \delta$ implies:

$$u(t) = u_o + K(u_1)e^{-\mu t} + \varepsilon(t)$$

where:

$$\|\varepsilon(t)\|_\alpha \leq C\,\|u_1 - u_o\|_\alpha\, e^{-\gamma t}.$$

Here $K(u_o) = 0$ and $K(\cdot)$ is a continuous map from a neighborhood of u_o in Y^α into the one-dimensional space $N(L - \beta I)$, and, if E_1 is the associated projection onto $N(L - \beta I)$,

$$K(u_1) = E_1(u_1 - u_o) + O(\|u_1 - u_o\|_\alpha^2) \quad \text{as} \quad u_1 \to u_o.$$

(iv) (Instability) Assume $\sigma(L) \cap \{\mathrm{Re}\,z < 0\}$ is a nonempty spectral set. Then the equilibrium solution u_o is unstable.

Application to the stability analysis of the trivial state $(\tilde{s}, \tilde{a})$. We verify the hypotheses of Theorem 6.2. Here A and $r'(u_o)$ are the operators L_o and λL_1 defined in (6.87), while Y and $D(L_o) = X$ are the spaces defined in (6.21), (6.22), and (6.23).

(i) We know that L_o is m-sectorial.

(ii) L_1 is a bounded linear map from Y^α to Y, for some α in $0 \leq \alpha < 1$. Indeed L_1 is bounded from Y to Y, and it is well known that Y^α is continuously imbedded into Y^β if $\alpha > \beta$, whence:

$$|L_1 U|_Y \leq C|U|_Y = C\,\|U\|_o \leq C\,\|U\|_\alpha.$$

(iii) r is Lipschitzian on Y^α. We apply again the continuous imbedding of Y^α into Y for $\alpha \geq 0$, and obtain:

$$|r(x) - r(y)|_Y \leq L|x-y|_Y \leq L\,\|x-y\|_\alpha$$

for every x and y in Y^α.

(iv) We finally have to verify the property (6.91) for:

$$g : \begin{bmatrix} u \\ v \end{bmatrix} \to \lambda\{R(\tilde{s}+u,\tilde{a}+v) - R(\tilde{s},\tilde{a}) - \tilde{R}_s u - \tilde{R}_a v\} \begin{bmatrix} 1 \\ 1 \end{bmatrix}.$$

The quantity within braces can be rewritten:

$$Q = \frac{1}{2} [R_{ss}(M)u^2 + 2R_{sa}(M)uv + R_{aa}(M)v^2]$$

where $M(x)$ is a point:

$$M(x) = (\tilde{s} + \theta(x)u(x), \tilde{a} + \theta(x)v(x)), \quad 0 < \theta(x) < 1.$$

Since R_{ss}, R_{sa}, and R_{aa} are bounded functions, we have:

$$Q(x) \leq C(u^2(x) + v^2(x)).$$

But we know [72,p.35] that we have the continuous inclusion, for $0 < \alpha < 1$,

$$Y^{\alpha} \subset C^{\nu}(\Omega) \times C^{\nu}(\Omega) \quad \text{if} \quad 0 \leq \nu < 2\alpha - \frac{n}{2}.$$

Therefore we have, if $\frac{n}{4} < \alpha < 1$,

$$|Q|_{L^2(\Omega)} \leq C_1(|u|_o^2 + |v|_o^2) \leq C_2 \, ||U||_{\alpha}^2$$

where $|u|_o = \max_{x\in\overline{\Omega}} |u(x)|$ and $U = (u,v)$.

Thus $|g(U)|_Y = O(\,||U||_{\alpha}^2\,)$.

But [72,p.35] we have the continuous inclusion for $0 < \alpha < 1$

$$Y^{\alpha} \subset H^1(\Omega) \times H^1(\Omega) \text{ if } \alpha > 1/2.$$

Thus, combining the conclusions of Theorem 6.2 with the discussion of Section 6.3, we obtain the following proposition, where $(c_n,d_n) = (\mu_n/z'', \mu_n/z')$ denotes, for $n = 1,2,\ldots$, the intervals defined in this section:

Proposition 6.2 - Suppose $\max(1/2,n/4) < \alpha < 1$. Then:

(i) (6.1) has an unique solution on $[0,\infty[$ with initial value $(s_1,a_1) \in Y^{\alpha}$. It is a continuous function $(s,a) : [0,\infty[\to Y$ such that $s(0) = s_1$, $a(0) = a_1$ and on $]0,+\infty[$ we have $(s(t),a(t)) \in V \times V$.

(ii) If λ lies outside $\bigcup_{n=1}^{\infty} [c_n,d_n]$, then $(\tilde{s},\tilde{a})$ is asymptotically stable. More precisely there exist $\delta > 0$ and $c > 0$ such that if:

$$\|(s_1-\tilde{s},\ a_1-\tilde{a})\|_{\alpha} < \delta,$$

then on $0 \leq t \leq \infty$ the solution of (6.1) satisfies:

$$\|(s(t) - \tilde{s},\ a(t) - \tilde{a})\|_{H^1(\Omega)\times H^1(\Omega)} \leq c e^{-\beta t} \|(s_1-\tilde{s},\ a_1-\tilde{a})\|_{\alpha}$$

where β is the least eigenvalue of L_{λ}.

(iii) Assume $\lambda \in \bigcup_{n=1}^{\infty}]c_n,d_n[$. Then the equilibrium solution $(\tilde{s},\tilde{a})$ is unstable.

Remark 6.1 - The reasonings of this section can be easily extended to the stability analysis of any stationary solution v of (6.1), even with spatially non uniform concentrations: The spectrum of $f_u(v)$ $(= L_o + r'(v))$ is only composed of isolated eigenvalues, and the stability of the steady state depends upon the sign of the real part of the least eigenvalue.

6.6.3 - f is a twice continuously differentiable mapping: For the mapping $f : X \to Y$ defined by (6.20)-(6.23), the first few derivatives are:

$$f_u(\lambda,u) : X \to Y,\ f_{\lambda}(\lambda,u) : R \to Y,\ f_{uu}(\lambda,u) : X \times X \to Y,$$

$$f_{u\lambda}(\lambda,u) : X \times R \to Y, \quad \text{and} \quad f_{\lambda\lambda}(\lambda,u) : R \times R \to Y$$

defined by:

$$f_u(\lambda,u) \begin{bmatrix} v \\ w \end{bmatrix} = \begin{bmatrix} -\Delta + \lambda(R_s+1) & \lambda R_a \\ \lambda R_s & -\beta\Delta + \lambda(R_a+\alpha) \end{bmatrix} \begin{bmatrix} v \\ w \end{bmatrix}$$

where: $R_s = \frac{\partial R}{\partial s}(s,a)$, ...,

$$f_\lambda(\lambda,u)\xi = \xi \begin{bmatrix} R(s,a) - (s_o - s) \\ R(s,a) - \alpha(a_o - a) \end{bmatrix},$$

$$f_{uu}(\lambda,u)\begin{bmatrix} v \\ w \end{bmatrix}\begin{bmatrix} y \\ z \end{bmatrix} = \lambda[R_{ss}\ vy + R_{sa}(vz + wy) + R_{aa}\ wz]\begin{bmatrix} 1 \\ 1 \end{bmatrix},$$

where: $R_{ss} = \frac{\partial^2 R}{\partial s^2}(s,a)$, ...,

$$f_{u\lambda}(\lambda,u)\begin{bmatrix} v \\ w \end{bmatrix}\xi = \xi \begin{bmatrix} R_s + 1 & R_a \\ R_s & R_a + \alpha \end{bmatrix}\begin{bmatrix} v \\ w \end{bmatrix}$$

and $f_{\lambda\lambda} = 0$.

Here R_{ss} for example denotes the mapping:

$R_{ss}(s(\cdot),a(\cdot)) : H^2(\Omega) \times H^2(\Omega) \to L^2(\Omega)$ which associates, to each pair of functions (v,y) in $H^2(\Omega) \times H^2(\Omega)$, the function $R_{ss}(s,a)$ $vy : x \to R_{xx}(s(x),a(x))v(x)y(x)$, $x \in \Omega$. This mapping is continuous. Indeed, since $H^2(\Omega)$ is continuously imbedded into $C_o(\bar{\Omega})$ for $n < 4$, we have:

$$|v(x)| \leq |v|_o \leq c_2 \|v\|_{H^2(\Omega)} \quad \text{and} \quad |y(x)| \leq |y|_o \leq c_2 \|y\|_{H^2(\Omega)}$$

and, since $|R_{ss}(\xi,\eta)|$ is bounded for $\xi \in R$ and $0 \leq \eta \leq a_o$, $|R(s(x),a(x))| < c_1$, whence:

$$\|R(s,a)vy\|_{L^2(\Omega)} \leq c\,\|v\|_{H^2(\Omega)}\,\|y\|_{H^2(\Omega)},$$

where c, c_1, c_2 are various constants.

Thus $R_{ss}(s,a) \in \mathcal{L}(X,L^2(\Omega))$, space of bounded linear operators from $X = V \times V$ into $L^2(\Omega)$. We now have to show the continuity of the mapping $R_{ss} : X \to \mathcal{L}(X,L^2(\Omega))$ associating to each pair of functions (s,a) in X the operator $R_{ss}(s,a)$. This property is an immediate consequence of the Lipschitz continuity of the function R_{ss}, which implies that:

$$|R_{ss}(s+h,\ a+k)vy - R_{ss}(s,a)vy|_{L^2(\Omega)} \leq c(|h|_{L^2(\Omega)} + |k|_{L^2(\Omega)})$$

$$\times\ \|v\|_{H^2(\Omega)}\ \|y\|_{H^2(\Omega)},$$

whence:

$$\|R_{ss}(s+h,\ a+k) - R_{ss}(s,a)\|_{L(X,L^2(\Omega))} \leq c(|h|_{L^2(\Omega)} + |k|_{L^2(\Omega)}).$$

6.6.4 - The range of $f_u(\lambda,u_o)$ is closed and of codimension 1. The following definitions and theorem can be found, together with the proof, in [76,77].

Definition 6.5 - Let X,Y be two Banach spaces. $T \in L(X,Y)$ (space of bounded linear operators from X into Y) is called a Fredholm operator if:

(i) N(T) is finite dimensional

(ii) Y/T(X) is finite dimensional.

T is said to be Fredholm of type (p,q) if N(T) (resp. Y/T(X)) is p-dimensional (resp. q-dimensional).

Let F(X,Y) denote the set of Fredholm operators from X into Y.

Definition 6.6 - The index function ind : $F(X,Y) \rightarrow Z$ is defined by ind(T) = dim N(T) - dim Y/T(X).

Theorem 6.3 -

(i) Let X,Y be two Banach spaces and $T \in F(X,Y)$. Then T(X) is closed in Y.

(ii) If $T \in F(X,Y)$ and $K \in K(X,Y)$ (space of compact linear operators from X into Y), then $T + K \in F(X,Y)$ and ind (T+K) = ind (T).

Application to $f_u(\lambda,u_o)$ defined in (6.86) - Here the spaces X and Y are those defined in (6.21)-(6.23). Since $f_u(\lambda,u_o) = T + K$ where

$$T = \begin{bmatrix} -\Delta + 1 & 0 \\ 0 & -\beta\Delta + \alpha \end{bmatrix} \text{ and } K = \lambda \begin{bmatrix} \tilde{R}_s & \tilde{R}_a \\ \tilde{R}_s & \tilde{R}_a \end{bmatrix}$$

all we have to check is that $T \in F(X,Y)$ with index 0 and $K \in K(X,Y)$. Since $-\Delta + 1$ (resp. $-\beta\Delta + \alpha$) is an isomorphism of V onto $L^2(\Omega)$, the former property is immediate. Since the imbedding of X into Y is compact and $K \in \mathcal{L}(Y)$, K is compact as the product of a bounded operator and a compact operator. We thus have:

Proposition 6.3 - The operator $f_u(\lambda,u_o)$ defined in (6.86) is Fredholm with index 0. Its range is closed. If λ is one of the endpoints of intervals (c_n,d_n), $n = 1,2,\ldots,$ then:

$$\dim N(f_u(\lambda,u_o)) = \operatorname{codim} R(f_u(\lambda,u_o)) = d$$

where d is the dimension of the null space of $-\Delta-\mu_n I$.

Remark 6.2 - In all the examples encountered we had d = 1. But d could be greater than 1. For example on the surface of a sphere the eigenfunctions are the spherical harmonics and the null-space associated to μ_n is 2n + 1-dimensional.

Conclusion - Diffusive instability is a somewhat unexpected phenomenon since it is commonly accepted that diffusion has a smoothing action and a tendancy to make uniform the values in a field of temperatures or of concentrations. The possible occurrence of diffusion driven instability in morphogenesis justifies the study of systems like the s-a system, which is only an example in a large class of 2-species systems involving diffusion, enzyme reaction, and transport from the outside. Pattern formation in fluids is known since a long time, with Bénard and Taylor phenomena. In ecological systems it is called patchiness, and we refer to Okubo's book [78] and the references therein for a detailed study of this phenomenon. Finally, for discussions about the relations between pattern formation by dynamic systems and pattern recognition, we refer to [79].

References

[1] Turing, A.M., The chemical basis of morphogenesis, Phil. Trans. Roy. Soc. London, Vol. B 237 (1952) 37-72.

[2] Gmitro, J.I. and Scriven, L.E., in: A Physicochemical Basis for Pattern & Rhythm, in: Warren, K.B. (ed.), Intracellular Transport (Academic Press, New York, 1966).

[3] Prigogine, I. and Nicolis, G., On symmetry breaking instabilities in dissipative systems, J. Chem. Phys., 46 (1967) 3542-3550.

[4] Prigogine I. and Lefever R., Symmetry breaking instabilities in dissipative systems II, J. Chem. Phys. 48 (1968) 1695-1700.

[5] Prigogine I., Lefever R., Goldbeter A., and Herschkowitz-Kaufman M., Symmetry breaking instabilities in biological systems, Nature, 223 (1969) 913-916.

[6] Goodwin, B.C. and Cohen, M.H., A phase shift model for the spatial and temporal organisation of developing systems, J. Theor. Biol., 25 (1969) 49-107.

[7] Othmer, H.G. and Scriven, L.E., Interactions of reaction and diffusion in open systems, I. & E. C. Fund. 8 (1969) 303-313.

[8] Glansdorff, P. and Prigogine, I., Thermodynamic Theory of Structure, Stability and Fluctuations (Wiley Interscience, New York, 1971).

[9] Othmer, H.G. and Scriven, L.E., Instability and dynamic pattern in cellular networks, J. Theor. Biol. (1971) 32, 507-537.

[10] Gierer, A. and Meinhardt, H., A theory of biological pattern formation, Kybernetika (Prague) 12 (1972) 30-39.

[11] Herschkowitz-Kaufmann, M. and Nicolis, G., Localized spatial structures and nonlinear chemical waves in dissipative systems, J. Chem. Phys. 56 (1972) 1890-1895.

[12] Boa, J.A., A model biochemical reaction, Ph.D. Thesis, Caltech, Pasadena (1974).

[13] Gierer, A. and Meinhardt, H., Biological pattern formation involving lateral inhibition: Levin, S.A. (ed.), Some Mathematical Questions in Biology 6 (The American Mathematical Society, Providence, 1974).

[14] Meinhardt, H. and Gierer, A., Applications of a theory of biological pattern formation based on lateral inhibition, J. Cell. Sci., 15 (1974) 321-346.

[15] Nicolis, G. and Auchmuty, J.F.G., Dissipative structures, catastrophes and pattern formation: A bifurcation analysis, Proc. Nat. Acad. Sci. 71 (1974) 2748-2751.

[16] Wolpert, L., Positional information and the development of pattern and form: Cowan, J.D. (ed.), Some Mathematical Questions in Biology 5 (The American Mathematical Society, Providence, 1974).

[17] Auchmuty, J.F.G. and Nicolis, G., Bifurcation analysis of non-linear reaction-diffusion equations. I : Evolution equations and the steady state solutions, Bull. Math. Biol. 37 (1975) 323-365.

[18] Babloyantz, A. and Hiernaux, J., Models for cell differentiation and generation of polarity in diffusion-controlled morphogenetic fields, Bull. Math. Biol. 37 (1975) 637-657.

[19] Boa, J.A., Multiple steady states in a model biochemical reaction, Studies in Appl. Math., Vol. 54 N°1 (1975).

[20] Herschkowitz-Kaufman , M., Bifurcation analysis of a nonlinear reaction diffusion equation. II : Steady state solutions and comparison with numerical simulations, Bull. Math. Biol. 37 (1975) 589-635.

[21] Maginu, K., Reaction-diffusion equations describing morphogenesis I : Waveform stability of stationary wave solutions in a one-dimensional model, Math. Biosciences 27 (1975) 17-98.

[22] Othmer, H.G., Nonlinear wave propagation in reacting systems, J. Math. Biol. 2 (1975) 133-163.

[23] Boa, J.A. and Cohen, D.S., Bifurcation of localized disturbances in a model biochemical reaction, SIAM J. Appl. Math., Vol. 30, N°1 (1976) 123-135.

[24] Fife, P.C., Pattern formation in reacting and diffusing systems, J. Chem. Phys. 64 (1976) 554-564.

[25] Goodwin, B.C., On some relationships between embryogenesis and cognition, Theoria to Theory, 10 (1976) 33-44.

[26] Goodwin, B.C., Analytical Physiology of Cells and Developing Organisms (Academic Press, New York, 1976).

[27] Babloyantz, A., Mathematical models for morphogenesis: Levin, S.A (ed.), Mathematical Models in Biological Discovery (Lecture Notes in Biomathematics N°13, Springer-Verlag, Berlin, 1977).

[28] Fife, P.C., Stationary patterns for reaction-diffusion equations: Nonlinear Diffusion, Res. Notes in Math. 14 (Pitman, London, 1977).

[29] Fife, P.C., On modelling pattern formation by activator-inhibitor systems, J. Math. Biol. 4 (1977) 358-362.

[30] Haken, H., Synergetics, An Introduction, Nonequilibrium Phase Transitions and Self-organization in Physics, Chemistry, and Biology (Springer-Verlag, Berlin, 1977).

[31] Kernevez, J.P., Bunow, B., Duban, M.C., Joly, G., and Thomas,D., Structuration en espace spontanée à l'intérieur d'une membrane monoenzymatique, C.R. Acad. Sci. Paris Ser. 1-B 284 (1977) N°3 A195-A198.

[32] Kernevez, J.P., Duban, M.C., Joly, G., and Thomas, D., Spatio-temporal structuration in immobilized enzyme systems: Perlmutter, A. and Scott, L.F. (eds.), The Significance of Nonlinearity in the Natural Sciences (Plenum, New York, 1977).

[33] Meinhardt, H., A model of pattern formation in insect embryogenesis, Journal Cell. Sci., 23 (1977) 117-139.

[34] Meurant, G. and Saut, J.C., Bifurcation and stability in a chemical system, J. Math. Anal. and Appl. 59 (1977) 69-91.

[35] Murray, J.D., Lectures on Nonlinear Differential-Equation Models in Biology (Clarendon Press, Oxford, 1977).

[36] Nicolis, G. and Prigogine, I., Self-Organization in Nonequilibrium Systems, from Dissipative Structures to Order through Fluctuations, (Wiley-Interscience, New York, 1977).

[37] Othmer, H.G., Current problems in pattern formation: Levin,S.A. (ed.), Some Mathematical Questions in Biology 8 (The American Mathematical Society, Providence, 1977).

[38] Ashkenazi, M. and Othmer, H.G., Spatial patterns in coupled biochemical reactors, J. Math. Biology 5 (1978) 305-350.

[39] Erneux, T., Hiernaux, J. and Nicolis, G., Turing's theory in morphogenesis, Bull. Math. Biol., 40 (1978) 771-789.

[40] Kauffman, S.A., Shymko, R.M., and Trabert, K., Control of sequential compartment formation in Drosophila, A uniform mechanism may control the locations of successive binary developmental commitments, Science, Vol. 199 (1978) 259-270.

[41] Mimura , M. and Murray, J.D., Spatial structures in a model substrate-inhibition diffusion system, Z. Naturfosch, 33C (1978) 580-586.

[42] Sijbrand, J., Pattern formation in biological systems: An analytical approach, Annals of Systems Research, 7 (1978) 73-89.

[43] Fife, P.C., Mathematical Aspects of Reacting and Diffusing Systems, (Springer-Verlag, Berlin, 1979).

[44] Herschkowitz-Kaufman, M. and Erneux, T., The bifurcation diagram of model chemical reactions, Annals of the New York Academy of Sciences, Vol. 316 (1979) 296-313.

[45] Hiernaux, J. and Erneux, T., Chemical patterns in circular morphogenetic fields, Bull. Math. Biol., 41 (1979) 461-468.

[46] Kernevez, J.P., Joly, G., Duban, M.C., Bunow, B., and Thomas,D., Hysteresis, oscillations, and pattern formation in realistic immobilized enzyme systems, J. Math. Biology 7 (1979) 41-56.

[47] Mimura, M. and Nishiura, Y., Spatial patterns for an interaction-diffusion equation in morphogenesis, J. Math. Biology 7 (1979) 243-263.

[48] Garcia-Bellido, A. and Merriam, J.P., Parameters of the wing imaginal disc development of Drosophila melanogaster, Develop. Biol., 24 (1971) 61-87.

[49] Garcia-Bellido, A, Ripoll, P., and Morata, G., Developmental compartmentalization in the wing disk of drosophila, Nature New Biol., 245, (1973) 251-253.

[50] Garcia-Bellido, A., in "Cell Patterning", Ciba Foundation Symp. 29 (Elsevier, Amsterdam, 1975).

[51] Garcia-Bellido, A., Ripoll, P., and Morata, G., Developmental compartmentalization in the dorsal mesothoracic disc of Drosophila, Develop. Biol. 48 (1976) 132-147.

[52] Sprey, T.E., Morphological and histochemical changes during the development of some of the imaginal disks of calliphora erythrocephala, Netherlands Journal of Zoology, 20 (2) (1970) 253-275.

[53] Sprey, T.E., Aldehyde oxidase distribution in the imaginal discs of some diptera, Wilhelm Roux's Archives 183 (1977) 1-75.

[54] Steiner, E., Establishment of compartments in the developing leg imaginal discs of Drosophila melanogaster, Wilhelm Roux's Archives 180 (1976) 9-30.

[55] Sprey, H.T., personal communication (1978).

[56] Child, C.M., Patterns and Problems of Development (The University of Chicago Press, Chicago, 1941).

[57] Webster, G., Morphogenesis and pattern formation in hydroids, Biol. Rev. 46 (1971) 1-46.

[58] Wall, R., Physiological gradients in development, A possible role for messenger ribonucleoprotein, Adv. in Morphogenesis 10 (1973) 41-114.

[59] Sander, K., Pattern specification in the insect embryo, in: "Cell Patterning", Ciba Foundation Symp. 29 (Elsevier, Amsterdam, 1975).

[60] Nieuwkoop, P.D., Origin and establishment of embryonic polar axis in amphibian development, Curr. Top. in Develop. Biol. 11 (1977) 115-132.

[61] Brenner, M., Cyclic AMP gradient in migrating pseudoplasmodia of the cellular slime mold Dictyostelium discoideum, J. Biol. Chem. 252 (1977) 4073-4077.

[62] Lovtrup, S., Landström, V., and Lovtrup-Rein, H., Polarities, Cell differentiation, and primary induction in the amphibian embryo, Biol. Rev. 53 (1978) 1-42.

[63] Keller, H.B., Numerical solution of bifurcation and nonlinear eigenvalue problems: Rabinowitz, P.H. (ed.), Applications of Bifurcation Theory, (Academic Press, New York, 1977).

[64] Keller, H.B., Two new bifurcation phenomena, IRIA Research Report N°369 (1979).

[65] Decker, D.W., Topics in bifurcation theory, Ph.D. Thesis, California Institute of Technology, Pasadena, California 1978 (152 pp.+V).

[66] Erneux, T. and Hiernaux, J., Transition from polar to duplicate patterns, (preprint).

[67] Bathe, K.J. and Wilson, E.L., Numerical Methods in Finite Element Analysis (Prentice-Hall, New Jersey, 1976).

[68] Strang, G. and Fix, G.J., An Analysis of the Finite Element Method, (Prentice Hall, Englewood Cliffs, 1973).

[69] Bunow, B., Kernevez, J.P., Joly, G., and Thomas, D., Pattern Formation by reaction diffusion instabilities: Application to Kauffman's model for morphogenesis in drosophila, to be published, J. Theor. Biol.

[70] Garcia-Bellido, A., Lawrence, P., and Morata, G., Sci. Am. 241 (1979) 102.

[71] Joly, G., Analyse mathématique de structurations spatiotemporelles dans des systèmes à enzymes immobilisées, Thesis, to appear.

[72] Henry, D., Geometric theory of semilinear parabolic equations, Lecture Notes, University of Kentucky, 1976.

[73] Kato, T., Perturbation Theory for Linear Operators (Springer-Verlag, New York, 1966).

[74] Iooss, G., Bifurcation et stabilité, Publications Mathématiques d'Orsay, N°31,(Université de Paris Sud, Orsay, 1972).

[75] Lions, J.L., Problèmes aux Limites dans les Equations aux Dérivées Partielles (Les Presses de l'Université de Montréal, Montréal, 1965).

[76] Palais, R., Lecture N°7, "Seminar on Atiyah-Singer index theorem", Princeton, 1965.

[77] Saut, J.C., Lecture N°1, "Séminaire équations aux dérivées partielles non linéaires, Publications Mathématiques d'Orsay, N°77-76, Orsay (1977).

[78] Okubo, A., Diffusion and Ecological Problems : Mathematical Models (Springer Verlag, Berlin, 1980).

[79] Haken, H. (ed.) Pattern Formation by Dynamic Systems and Pattern Recognition (Springer Verlag, Berlin, 1979).

CHAPTER 7

OSCILLATIONS AND WAVE FRONT PROPAGATION

In Chapters II to VI we mainly dealt with stationary solutions of enzyme distributed systems, our goal being to present, in Chapters V and VI, spatial structurations in the s-a system. In the present chapter we describe other types of structurations, namely spatio-temporal structurations such as oscillations and wave front propagation. In Section 7.1 we talk about oscillations in the so-called papain system, in Section 7.2 we give an account of oscillations in a two substrates system catalyzed by uricase, and in Section 7.3 we show that the s-a system of Chapter 6 may propagate waves, for parameter values suited to this purpose. We only give numerical results, and omit both mathematical and numerical aspects including Hopf bifurcation and continuation of periodic solutions. Indeed these subjects would necessitate another book!

From the biological point of view, sustained oscillations have been experimentally observed and established beyond doubt for glycolysis [1,2,4]. Models to represent the observed oscillations have been worked out [7,2]. A model taking explicitly into account the allosteric effects which is free from phenomenological factors has been constructed [14]. Circadian rhythms exhibit approximately 24-hour periodicity. General accounts of oscillatory phenomena are given in [3,16]. [29] is devoted more specifically to oscillatory enzymatic reactions. We have listed from [1] to [39] in chronological order, a sample of the experimental and theoretical work on biological, biochemical, and chemical oscillations. The two "stars" are the glycolytic oscillation and the Belousov-Zhabotinskii oscillation . However the latter is not a biochemical oscillation , but occurs during the cerium ion catalyzed oxidation of malonic acid by bromate ion. This reaction requires elaborate kinetics to account for its observed behavior [27].

$$R = a \frac{H}{1 + bH + cH^2}$$

Thus papain has an autocatalytic character in hydrogen-ion concentration, and is inhibited by excessive H.

For H held fixed, the dependence of R upon S is Michaelian, that is of the form:

$$R = V_M \frac{S}{K_S + dS} \quad .$$

Thus H and S in the papain system play the role of S and A in the S-A system.

We may consider [32] that equilibrium between the combination of hydrogen and hydroxyl ions (concentration OH) to form water, and the reverse ionization of water, is instantaneous as compared with the enzyme reaction rate:

$$K_W = H \cdot OH = 10^{-20} \text{ mole}^2\text{cm}^{-6}. \tag{7.3}$$

7.1.2 - <u>Papain kinetics in a membrane</u>: A membrane (thickness L) bearing papain is formed on a glass pH electrode and immersed in a bath of substrate. [22]. The conservation equations governing the system may be formulated as:

$$(7.4)\quad \begin{cases} S_t - D_S S_{xx} + R(S,H) = 0, & 0 < x < L, \quad t > 0, \\ A_t - D_H A_{xx} - R(S,H) = 0, & \\ A = H - K_W/H, & \\ S(0,t) = S_o, \quad A(0,t) = A_o, & \\ S_x(1,t) = A_x(1,t) = 0 & \end{cases}$$

where R(S,H) is given by (7.1), together with (7.2), and D_S and D_H are the diffusion coefficients of S and H within the membrane. The equation for A = H - OH is obtained by using (7.3) and subtracting the equation for OH^- from the equation for H^+:

$$H_t = D_H H_{xx} - R(S,H) - \frac{\partial}{\partial t}(H_2O), \qquad \text{and}$$

$$(OH)_t = D_H (OH)_{xx} - \frac{\partial}{\partial t} (H_2O)$$

Here we assume the same diffusivity for hydrogen and hydroxyl ions, and take into account the rate of water formation.

S_o is the bath substrate concentration, and $A_o = P_o - K_W/P_o$ where P_o is the bath H^+ concentration, more precisely:

$$P_o = 10^{-13} \text{ mole cm}^{-3}.$$

Thus the external pH in the bath is kept at pH = 10.

For $L = 10^{-2}$cm, $S_o = 5.5 \cdot 10^{-7}$ mole cm^{-3}, $E_o = 10^{-6}$ mole cm^{-3}, $D_H = 6.4 \times 10^{-6}$ cm^{2}sec^{-1}, and $D_S = 1.3 \times 10^{-6}$ cm^{2}sec^{-1}, an oscillation in time of the measured pH inside the membrane spontaneously occurs [22]. The period of oscillation is around 1/2 minute. This membrane has been studied theoretically [15,19]. There is qualitative agreement between the computer simulation and the experimental results.

7.1.3 - Numerical methods: Approximate solutions to (7.4) can be obtained by using either explicit or implicit finite difference schemes. Whether the step length is taken to be L/20 or L/100, both methods give similar results. It is somewhat unexpected and rather a good surprise that even the simple explicit method with a reasonable time step give good results, because, as will be seen, the values of H range over 6 decades, from 10^{-13} to 10^{-7}.

Due to this variation of H over 6 decades, special care must be exercised for the solution of the equation:

$$A = H - 10^{-20}/H, \quad \text{or} \quad H^2 - AH - 10^{-20} = 0,$$

where A is given. Instead of calculating the positive root by the formula:

$$H = 0.5 \; (A + (1 + 4.10^{-20})^{1/2}),$$

it is preferable to employ fixed point algorithms such as:

$$H_{n+1} = A + 10^{-20}/H_n \quad \text{if } A > 0, \qquad \text{or}$$

leading eigenvalues crosses zero. We thus have a Hopf bifurcation, and it was verified numerically [33] that the system (7.5) oscillates for every $\sigma \in (\sigma_1, \sigma_2)$. Figure 7.2 shows in the complex plane the two complex conjugate leading eigenvalues of L_σ as σ varies.

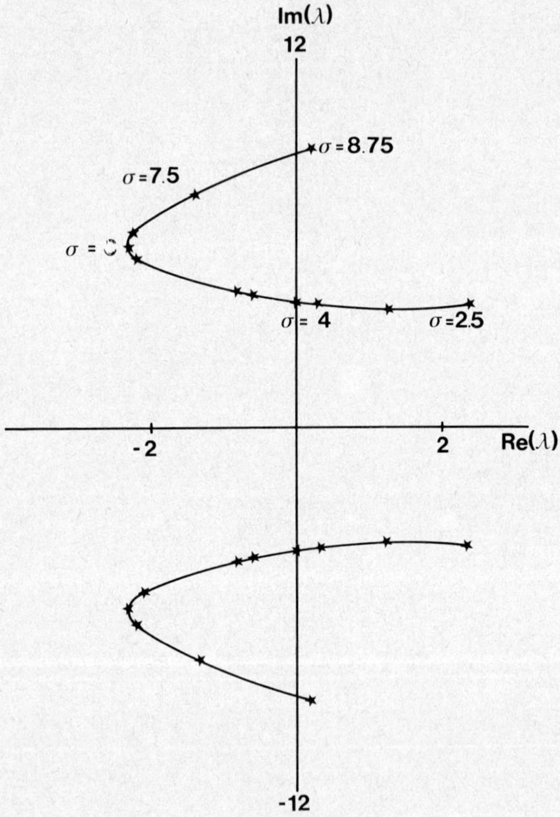

Fig. 7.2

It is not evident whether the Hopf bifurcation at $\sigma = \sigma_1$ for example is forward or backward. One could have a forward bifurcation of stable oscillations (s.o.) from a branch of stable steady states (s.s.s.) which become unstable (u.s.s.), as indicated schematically in Figure 7.3 (top); or a backward bifurcation of unstable periodic solutions, followed by a turning point and a branch of stable periodic solutions, as indicated in Figure 7.3 (bottom). Which eventuality actually occurs cannot be recognized by simply solving the evolution equations (7.5) in the vicinity of σ_1. In both cases one passes, when solving equations (7.5) for successive values of σ, from a stable steady state to a stable periodic solution.

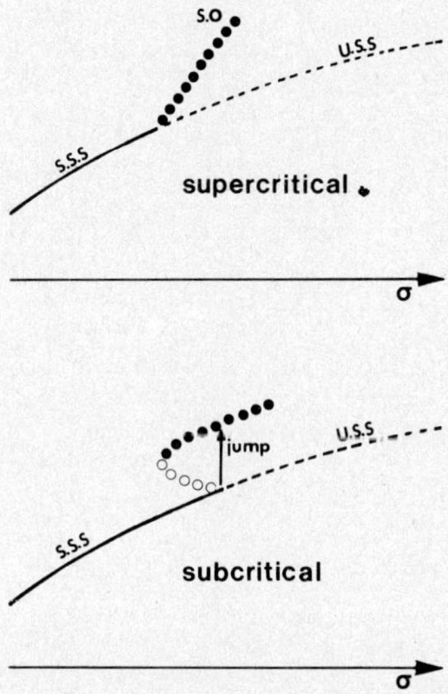

Fig. 7.3

Thus a numerical continuation of the curve of periodic solutions after the Hopf bifurcation would be helpful to provide more information about what actually occurs in the vicinity of σ_1 and σ_2 and between σ_1 and σ_2. Unfortunately numerical methods for detecting the bifurcation of periodic solutions are more complicated than for detecting the bifurcation of stationary solutions. This is mainly due to the fact that Hopf bifurcations do not correspond to the vanishing of the Jacobian determinant, but only to the vanishing of the real part of two complex conjugate eigenvalues. Moreover, the following of a curve of periodic solutions, after discretization with respect to space and time, involves much more degrees of freedom than for a curve of stationary solutions. For example for the problem (7.5), if we divide the space interval (0,1) into 20 equal intervals and the time interval (0,T), where T is the period, into 100 intervals, the "state" of the system is a vector in R^{2000}!

7.3 Wave Front Propagation in the s-a System

The system is governed by equations:

$$(7.8)\quad \begin{cases} s_t - s_{xx} + \lambda[R(s,a) - (s_o - s)] = 0, & 0 < x < 1, \quad t > 0, \\ a_t - \beta a_{xx} + \lambda[R(s,a) - \alpha(a_o - a)] = 0, & \\ s_x(0,t) = s_x(1,t) = a_x(0,t) = a_x(1,t) = 0, & \\ R(s,a) = \rho as/(1 + s + ks^2). & \end{cases}$$

Here the parameters s_o, a_o, α, ρ and k are such that, in the (s,a) phase plane (Figure 7.4) the isocline curves Γ_1 and Γ_2:

$$\begin{cases} R(s,a) - (s_o - s) = 0 & (\Gamma_1) \\ \text{and} & \\ R(s,a) - \alpha(a_o - a) = 0 & (\Gamma_2) \end{cases}$$

intersect at a point $(\tilde{s},\tilde{a})$ which is a stable steady state for the dynamical system:

$$(7.9)\quad \begin{cases} \dfrac{ds}{dt} + \lambda[R(s,a) - (s_o - s)] = 0 \\ \dfrac{da}{dt} + \lambda[R(s,a) - \alpha(a_o - a)] = 0. \end{cases}$$

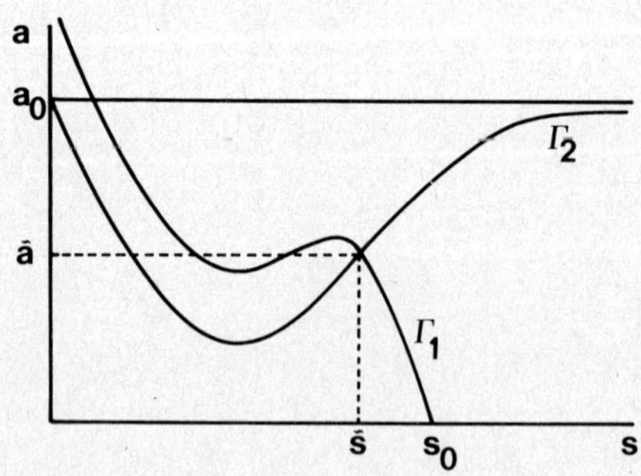

Fig. 7.4

Moreover, $(\tilde{s},\tilde{a})$ lies slightly on the right of the superior turning point of the first isocline Γ_1, as indicated in Figure 7.4.

An appropriate set of parameters for this to occur is:

(7.10) $k = 1$, $\rho = 0.9$, $s_o = 39.2$, $a_o = 784$, and $\alpha = 0.05$,

for which $\tilde{s} = 22$ and $\tilde{a} = 440$ [64].

It is well known that a particle, subject to the dynamics (7.9), and displaced from the equilibrium point $(\tilde{s},\tilde{a})$, can describe a large trajectory like the one in Figure 7.5 before coming back to the resting position $(\tilde{s},\tilde{a})$. In fact the direction of its movement is determined by the signs of $\frac{ds}{dt}$ and $\frac{da}{dt}$, which can change only by crossing the isocline curves, and are, for example, positive in the region near (0,0).

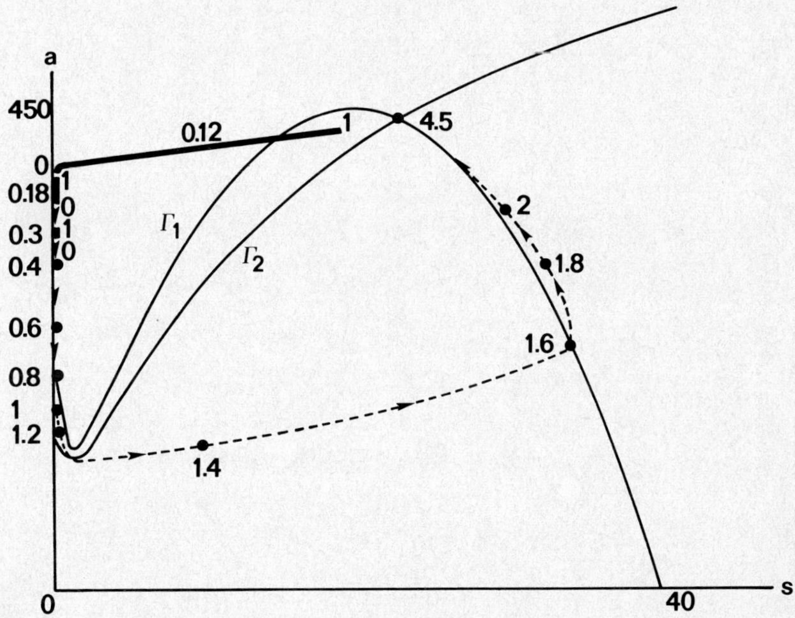

Fig. 7.5

Let us now turn our attention to the distributed system (7.8). We choose $\beta = 1$ and $\lambda \in (20,100)$, for which values, together with (7.10), the uniform steady state $(\tilde{s},\tilde{a})$ is stable. For this distributed system, the particle has to be replaced by a "snake" in the phase plane. At each time t the state of the system is a curve $[(s(x,t), a(x,t)), 0 < x < 1]$, which we shall call a "snake". The system being

$s(0,t) = 0.01$ for $0 < t < 0.1$, $s_x(0,t) = 0$ for $t > 0.1$,

$s_x(1,t) = 0$, $s(x,0) = \tilde{s}$

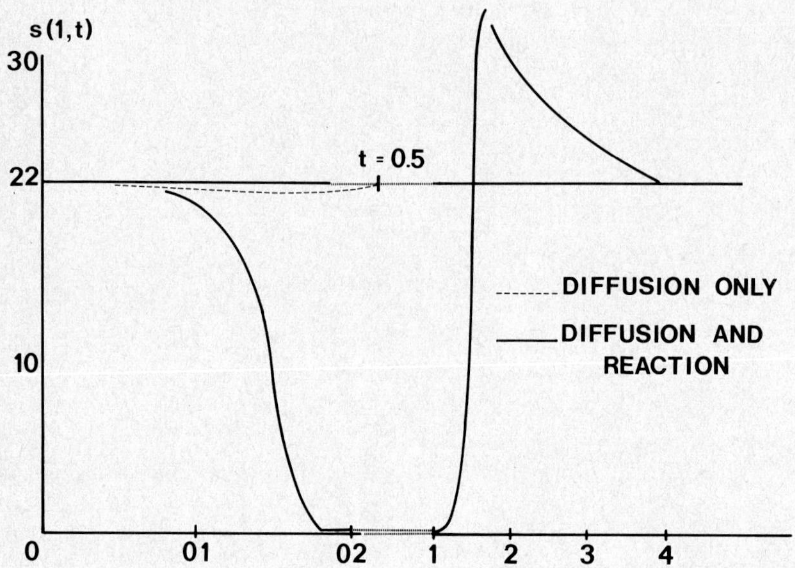

Fig. 7.7

We see from the responses s(1,t) of both systems that diffusion alone is unable to transmit from one end to another a wave front created by an excitation limited in time.

There are analogies between the behavior of our system and propagation of nervous impulse. In both cases there is a threshold, signal propagation, and, in the recovery phase, an overshoot.

However our model does not seem to be in competition with well established models of nerve impulse propagation, but maybe could aid to understand the dissemination of information in systems like slime molds [24,65].

Finally, Figure 7.8 shows the response s(1,t) of the distributed system (7.8) to a perturbation at x = 0 of duration $t_o = 0.05$, for $\lambda = 30$.

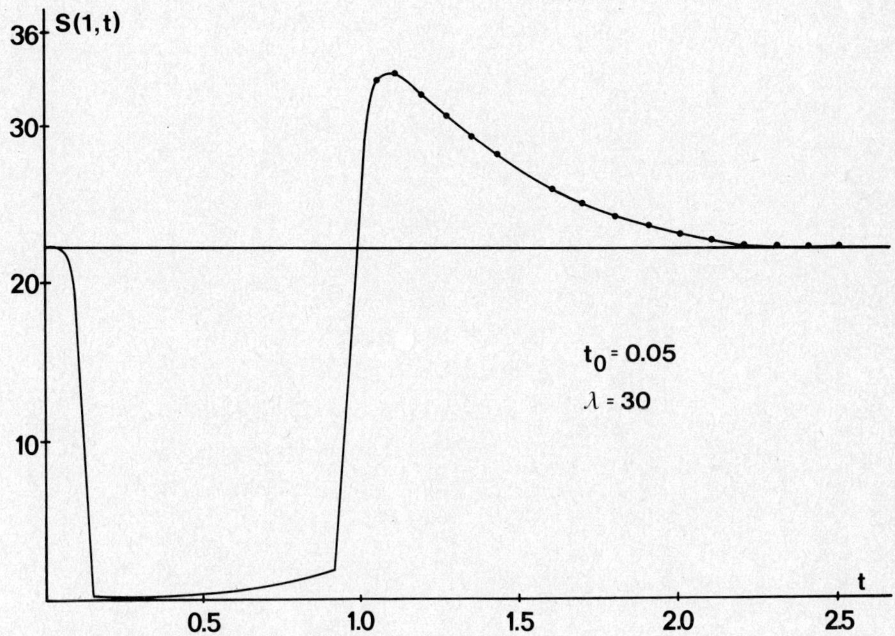

Fig. 7.8

Remark 7.1 - The biochemical interpretation of the above numerical results is that oscillations as well as wave front propagation can occur in immobilized enzyme systems, as a consequence of competing diffusion and enzyme reaction. In particular it is not necessary, for oscillations to occur, that several enzyme reactions be implied, with a possible feedback of the last product on the first reaction. A single reaction is sufficient, provided it is coupled to diffusion. Similarly the coupling of diffusion and enzyme reaction may induce the production of propagating wave fronts, much faster than purely diffusing chemical species.

References

[1] Hess, B., in: Functionelle und Morphologische Organisation der Zelle (Springer Verlag, Berlin, 1962).

[2] Higgins, J., A chemical mechanism for oscillations of glycolytic intermediates in yeast cells, Proc. Nat. Acad. Sci. U.S.A, 51 (1964) 989.

[3] Aschoff, J. (ed.), Circadian Clocks (North Holland, Amsterdam, 1965).

[4] Betz, A. and Chance, B., Phase relationship of glycolytic intermediates in yeast cells with oscillatory metabolic control, Arch. Biochem. Biophys. 109 (1965) 579-594.

[5] Gmitro, J.I. and Scriven, L.E., A physicochemical basis for pattern and rhythm: Warren, K.B. (ed.) Intracellular Transport, (Academic Press, New York, 1966).

[6] Winfree, A.T., Biological rhythms and the behavior of populations of coupled oscillators, J. Theor. Biol. 16 (1967) 15-42.

[7] Selkov, E.E., Self-oscillations in glycolysis, Europ. J. Biochem. 4 (1968) 79-86.

[8] Othmer, H.G. and Scriven, L.E., Interactions of reaction and diffusion in open systems, I. & E.C. Fund. 8 (1969) 303-313.

[9] Deshcherevskii, V.I., Zhabotinskii, A.M., Selkov, E.E., Sidorenko N.P., and Schnol, S.E. Oscillatory biological processes at the molecular level, Biofizika 15 (1970) 225-234.

[10] Selkov, E.E., Two alternative, self oscillating stationary states in thiol metabolism - Two alternative types of cell division, normal and malignant ones, Biofizika 15 (1970) 1065-1073.

[11] Selkov, E.E., Stable circadian rhythms as a property of cell populations, Life Sci. Space Res. 8 (1970) 157-167.

[12] Zaikin, A.N. and Zhabotinsky, A.M., Concentration wave propagation in two-dimensional liquid-phase self-oscillating system, Nature 225 (1970) 535-537.

[13] Othmer, H.G. and Scriven, L.E., Instability and dynamic pattern in cellular networks, J. Theor. Biol. (1971) 32, 507-537.

[14] Goldbeter, A. and Lefever, R., Dissipative structures for an allosteric model, Application to glycolytic oscillations, Biophys. J., 12 (1972) 1302-1315.

[15] Caplan, S.R., Naparstek, A., and Zabusky, N.J., Chemical oscillations in a membrane, Nature 245 (1973) 364.

[16] Chance, B., Pye, E.K., Ghosh, A.K., and Hess, B., Biological and Biochemical Oscillators (Academic Press, New York, 1973).

[17] Selkov, E.E., Analysis of temporal organisation of multienzyme systems, Abstr. Int. Biophys. Cong., Puschino (1973) 461-475.

[18] Pavlidis, T., Biological oscillators: Their Mathematical Analysis (Academic Press, New York, 1973).

[19] Zabusky, N.J. and Hardin, R.H., Phase transition, stability, and oscillations for an autocatalytic, single, first-order reaction in a membrane, Physical Review Letters, Vol. 31 N°13 (1973).

[20] Zhabotinsky, A.M. and Zaikin, A.N., Autowave processes in a distributed chemical system, J. Theor. Biol. 40 (1973) 45-61.

[21] Nicolis, G., Patterns of spatiotemporal organization in chemical and biochemical kinetics, SIAM-AMS Proc. 8 (1974) 33-58.

[22] Naparstek, A., Thomas, D., and Caplan, S.R., An experimental enzyme-membrane oscillator, Biochim. Biophys. Acta, 323 (1973) 643-646.

[23] Graves, D. in: Wingard, L.B. and Pye, K. (eds.) Enzyme Engineering N°2 (Plenum Press, New York, 1974).

[24] Gerisch, G. and Hess, B., Cyclic AMP controlled oscillation in suspended dictyostelium cells: Their relation to morphogenetic cell interactions, Proc. Nat. Acad. Sci. U.S.A., 71 (1974) 2118-2122.

[25] Hahn, H.S., Nitzan, A., Ortoleva, P., and Ross, J., Threshold excitations, relaxation oscillations, and effect of noise in an enzyme reaction, Poc. Nat. Acad. Sci. U.S.A., Vol. 71, N°10 (1974) 4067-4071.

[26] Field, R.J. and Noyes, R.M., Oscillations in chemical systems IV. Limit cycle behavior in a model of a real chemical reaction, J. Chem. Phys. Vol. 60, N°5 (1974) 1877-1884.

[27] Noyes, R.M. and Field, R.J., Oscillatory chemical reactions, Ann. Rev. Phys. Chem. 25 (1974) 95-119.

[28] Hastings, S.P. and Murray, J.D., The existence of oscillatory solutions in the Field-Noyes model for the Belousov-Zhabotinskii reaction, SIAM J. Appl. Math., 28, 3 (1975) 678-688.

[29] Goldbeter, A. and Caplan, S.R., Oscillatory enzymes, Ann. Rev. Biophys. Bioeng. 5 (1976) 449-476.

[30] Thames, H.D. Jr. and Elster, A.D., Equilibrium states and oscillations for localized two-enzyme kinetics: A model for circadian rhythms, J. Theor. Biol., 59 (1976) 415-427.

[31] Tyson, J.J., The Belousov-Zhabotinskii reaction, Lecture Notes in Biomathematics, Vol. 10, Springer Verlag, Berlin (1976).

[32] Hardt, S., Naparstek, A., Segel, L.A., and Caplan, S.R., Spatiotemporal structure formation and signal propagation in a homogeneous enzymatic membrane: Thomas, D. and Kernevez, J.P. (eds.) Analysis and Control of Immobilized Enzyme Systems (North Holland Amsterdam, 1976).

[33] Duban, M.C., Hysteresis, oscillations et structurations en espace dans des systèmes biochimiques distribués, Thèse de 3° Cycle: Contrôle des Systèmes: Compiègne (1976) 83 p.

[34] Auchmuty, J.F.G. and Nicolis, G., Bifurcation analysis of reaction-diffusion equations III Chemical oscillations, Bull. Math. Biol. 38 (1976) 325-350.

[35] Kauffman, S., Dynamic models of the mitotic cycle: Evidence for a limit cycle oscillator: Levin, S.A. (ed.) Some Mathematical Questions in Biology, 8 (The American Mathematical Society, Providence, 1977).

[36] Meurant, G. and Saut, J.C., Bifurcation and stability in a chemical system, J. Math. Anal. and Appl. 59 (1977) 69-91.

[37] Auchmuty, J.F.G., Qualitative effects of diffusion in chemical systems: Lectures on Mathematics in the Life Science, Vol. 10 (The American Mathematical Society, Providence, 1978).

[38] Ashkenazi, M. and Othmer, H.G., Spatial patterns in coupled biochemical reactors, J. Math. Biology 5, (1978) 305-350.

[39] Herschkowitz-Kaufman, M. and Erneux, T., The bifurcation diagram of model chemical reactions, Annals of the New York Academy of Sciences, Vol. 316 (1979) 296-313.

[40] Iooss, G., Bifurcation et stabilité, Lecture Notes N°31, Université de Paris XI (1973).

[41] Crandall, M.G. and Rabinowitz, P.H., The Hopf bifurcation theorem, MRC Technical Summary Report N°1604, University of Wisconsin, (1976).

[42] Marsden, J.E. and Mac Cracken, M, The Hopf bifurcation and its applications, Applied Mathematical Sciences 19, Springer Verlag, New York (1976).

[43] Nagumo, J., Arimoto, S., and Yoshizawa, S., An active pulse transmission line simulating nerve axon, Proc. Inst. Radio Engrs. 50 (1962) 2061-2070.

[44] Ortoleva, P.J. and Ross, J., Penetration of boundary perturbations in unstable chemical systems, The J. Chem. Phys., Vol. 56 N°1 (1972) 287-292.

[45] Herschkowitz-Kaufman,M. and Nicolis, G., Localized spatial structures and nonlinear chemical waves in dissipative systems, J. Chem. Phys., 56 (1972) 1890-1895.

[46] Rinzel, J. and Keller, J.B., Traveling wave solutions of a wave conduction equation, Bio. Jour. 13 (1973) 1313-1337.

[47] Winfree, A.T., Scroll-shaped waves of chemical activity in three dimensions, Science, 181 (1973) 937-939.

[48] Koppel, N. and Howard, L.N., Plane wave solutions to reaction-diffusion equations, Stud. Appl. Math., Vol. 52, N°4 (1973) 291-328.

[49] Howard, L.N. and Kopell, N., Wave trains, sharp fronts, and transition layers in reaction-diffusion equations, SIAM-AMS Proc. 8 (1974)

[50] Othmer, H.G., Nonlinear wave propagation in reacting systems, J. Math. Biol. 2 (1975) 133-163.

[51] Ortoleva, P.J. and Ross, J., Theory of propagation of discontinuities in kinetic systems with multiple time scales: Fronts,

Front multiplicity, and pulses, The J. Chem. Phys., Vol. 63, N°8 (1975) 3398-3408.

[52] Fife, P.C. and Mc Leod, J.B., The approach of solutions of non-linear diffusion equations to travelling wave solutions, Bull. Am. Math. Soc., 81, 6 (1975) 1076-1078.

[53] Rinzel, J., Neutrally stable traveling wave solutions of nerve conduction equations, Journal of Mathematical Biology 2, (1975) 205-217.

[54] Rinzel, J., Simple model equations for active nerve conduction and passive neuronal integration: Levin, S.A. (ed.) Some Mathematical Questions in Biology, 7 (The American Mathematical Society, Providence, 1976).

[55] Rinzel, J., Repetitive activity and Hopf bifurcation under point stimulation for a simple Fitzhugh-Nagumo nerve conduction model, J. Math. Biol. 5 (1978) 363-382.

[56] Fife, P.C., Singular perturbations and wave front techniques in reaction-diffusion problems, SIAM-AMS Proceedings, Vol. 10 (1976) 23-50.

[57] Cohen, H., Mathematical developments in Hodgkin-Huxley theory and its approximations: Levin, S.A. (ed.) Some Mathematical Questions in Biology, 7 (The American Mathematical Society, Providence, 1976).

[58] Sattinger, D.H., On the stability of waves of nonlinear parabolic systems, Advances in Mathematics, 22 (1976) 312-355.

[59] Sattinger, D.H., Weighted norms for the stability of traveling waves, Journal of Differential Equations 25 (1977) 130-144.

[60] Feinn, D. and Ortoleva, P., Catastrophe and propagation in chemical reactions, The Journal of Chemical Physics, Vol. 67, N°5 (1977) 2119-2131.

[61] Fife, P.C., Asymptotic analysis of reaction-diffusion wave fronts, Rocky Mountain J. Math. 7 (1977) 389-415.

[62] Fife, P.C. and Peletier, L.A., Nonlinear diffusion in population genetics, Arch. for Rat. Mec. and Anal., Vol. 64, N°2, (1977) 93-109.

[63] Cohen, D.S., Neu, J.C., and Rosales, R.R., Rotating spiral wave solutions of reaction-diffusion equations, SIAM J. Appl. Math. Vol. 35, N°3 (1978) 536-547.

[64] Kernevez, J.P., Murray, J.D., Joly, G., Duban, M.C., and Thomas D., Propagation d'onde dans un système à enzyme immobilisée, CRAS 287, A (1978) 961-964.

[65] Goldbeter, A., Erneux, T., and Segel, L.A., Excitability in the adenylate cyclase reaction in Dictyostelium Discoideum, FEBS Letters, 89, 2 (1978) 237-241.

[66] Fife, P.C., Asymptotic states for equations of reaction and diffusion, Bull. Am. Math. Soc. 84, 5 (1978) 693-726.

[67] Fife, P.C., Mathematical Aspects of Reacting and Diffusing Systems (Springer Verlag, Berlin, 1979).

[68] Britton, N.F. and Murray, J.D., Threshold, wave, and cell-cell avalanche behavior in a class of substrate inhibition oscillators, J. Theor. Biol., 77 (1979) 317-332.

[69] Carpenter, G.A., Bursting phenomena in excitable membranes, SIAM J. Appl. Math., Vol. 36, N°2 (1979) 334-372.

[70] Field, R.J. and Troy, W.C., The existence of solitary traveling wave solutions of a model of the Belousov-Zhabotinskii reaction, SIAM J. Appl. Math., Vol. 37, N°3 (1979) 561-587.